Grade **5**

KUMON MATH WORKBOOKS

Word Problems

P9-CRE-418

Table of Contents

KUMON

1   Pat at the toy factory is sorting toy cars into boxes. If he has 252 toy cars, and he wants to put 6 in each box, how many boxes does he need?   10 points

⟨Ans.⟩ _____

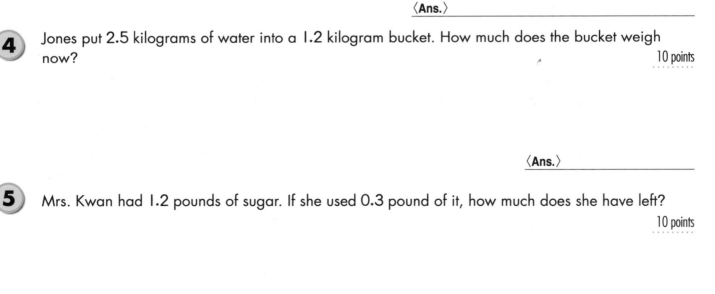

2   You bought 4 bananas and the total cost was $5. How much was 1 banana?   10 points

⟨Ans.⟩ _____

3   Ren is sorting his comic books. He has 68 comic books. If he puts 14 together as 1 set, how many sets will he have? How many comic books will remain?   10 points

⟨Ans.⟩ _____

4   Jones put 2.5 kilograms of water into a 1.2 kilogram bucket. How much does the bucket weigh now?   10 points

⟨Ans.⟩ _____

5   Mrs. Kwan had 1.2 pounds of sugar. If she used 0.3 pound of it, how much does she have left?   10 points

⟨Ans.⟩ _____

6 The sushi restaurant has 120 liters of soy sauce in the storage room. If they split that up into 3 containers, how many liters will be in each container?

10 points

〈Ans.〉

7 Lee has 24 songs on his phone so far, but Lucy only has 8. How many times more songs does Lee have on his phone than Lucy?

10 points

〈Ans.〉

8 Christina is trying to find her lost cat. If she divides up 260 flyers and gives 35 flyers to each person, how many people will get flyers, and how many flyers will she have left?

10 points

〈Ans.〉

9 Betty bought 6 pencils that cost 70¢ each and she paid $5. How much change did she get?   10 points

〈Ans.〉

10 Aki bought 1 comic book and 2 novels and paid $22. Her sister bought the same comic book and 4 novels and paid $38. If all the novels were the same price, how much was each novel?

10 points

〈Ans.〉

You remember this, right? Good!

Review

2

Date   /   /

Name

Level ☆

Score

/100

1 Debby wants to be the class president. She divided 150 flyers among her 5 helpers. How many flyers did each helper get?

10 points

⟨Ans.⟩ _____

2 In Carla's fridge, she has 1.5 liters of orange juice and 1.2 liters of apple juice. How much juice does she have?

10 points

⟨Ans.⟩ _____

3 At my summer camp, there are 6 basketball teams with 5 people on each team. If the camp counselors divide up 90 t-shirts equally amoug all of the basketball players, how many t-shirts will each basketball player get? Write one formula to solve the problem.

10 points

⟨Ans.⟩ _____

4 Football is more popular in Mike's town, so football card packs cost $6 at his local store, while baseball card packs only cost $4. If he bought 5 baseball card packs, and 3 football card packs, how much money did he spend? Write one formula to solve the problem.

10 points

⟨Ans.⟩ _____

5 The Henry family is wrapping presents. Mrs. Henry divides 90 centimeters of tape up, 7 centimeters at a time. How many pieces of tape did she make, and how much was left?

10 points

⟨Ans.⟩ _____

6 The delivery man has 68 packages to deliver. If he can carry 5 packages at a time, how many trips will it take him to deliver all his packages?

10 points

〈Ans.〉 _____

7 There were 2.5 liters of milk in the fridge. Then Hanna's mother used 1.2 liters to make an ice cream cake. How much milk is left?

10 points

〈Ans.〉 _____

 8 There are 365 days in a year. How many weeks are there in a year? How many days remain?

10 points

〈Ans.〉 _____

9 James wants to buy some ice creams for his friends. If one costs 80¢, and he has $5, how many ice creams can he buy, and how much money will he have left?

10 points

〈Ans.〉 _____

10 Alexis was looking at a set that included 1 hair clip that cost 25¢ and 1 hairpin that cost 15¢. If she paid 200¢, how many sets did she get?

10 points

〈Ans.〉 _____

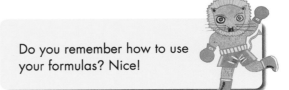

Do you remember how to use your formulas? Nice!

5

Level ☆☆

Date   /   /

Name

Score   /100

1 The local university football team drew 24,078 fans to their game last week. This week, 26,279 people came to the game. About how many people came to the 2 games in all? Round to the nearest ten-thousands place before adding.   10 points

Estimate of 24,078   Estimate of 26,279   Sum of estimates

$$\boxed{20,000} + \boxed{30,000} = \boxed{50,000}$$

⟨**Ans.**⟩ 50,000 people

Round to the nearest ten-thousands place before calculating!

2 In Atlanta, the baseball team might make the playoffs this year. 32,158 people came yesterday to see them play, and today 28,564 people are in the stadium. Round to the nearest ten-thousands place to estimate how many people came the last two days.   10 points

Estimate of 32,158   Estimate of 28,564   Sum of estimates

$$\boxed{} + \boxed{} = \boxed{}$$

⟨**Ans.**⟩ _____

3 There are 126,753 men and 114,397 women in Michele's city. About how many people are in her city? Round to the nearest ten-thousands place before adding.   10 points

⟨**Ans.**⟩ _____

4 Mr. Patel just got a big bonus. He went out and bought a riding lawnmower for $37,800 and a new car for $56,000. About how much did he spend altogether? Estimate to the nearest ten-thousands place.   10 points

⟨**Ans.**⟩ _____

5 At Brenda's new job, they bought a printer for $1,250 and a computer for $2,860. About how much did they spend in all? Estimate to the nearest thousands place.   10 points

⟨**Ans.**⟩ _____

6 Mesa's father is a salesman and he drove 36,020 kilometers last year. This year, he had to do more sales calls and drove 42,865 kilometers. About how many kilometers did he drive in all? Estimate to the nearest thousands place.   10 points

$$36,000 + 43,000 =$$

⟨Ans.⟩ _____

7 The town's largest supermarket chain earned $415,670 on May 1, but only $382,850 on May 2. About how much did they earn over the first two days of May? Calculate to the nearest thousand.   10 points

⟨Ans.⟩ _____

8 There are 124,310 square feet of cornfields around Ken's town. If there are also 268,170 square feet of wheat fields, about how many square feet of cropland is around his town? Calculate to the nearest thousand.   10 points

⟨Ans.⟩ _____

9 A soda plant bottled 18,356 bottles of soda yesterday and 21,695 today. About how many bottles of soda did they produce in the last two days? Estimate to the nearest thousand.   10 points

⟨Ans.⟩ _____

10 Kim's father jogs every day. He jogged 26,800 yards last week and 2,860 yards so far this week. About how much did he run in all? Estimate to the nearest thousand.   10 points

⟨Ans.⟩ _____

No problem, right? Good!

Round Numbers

4

Date / / 　 Name

Score /100

---

1 Last week the soccer team drew 32,407 fans, but this week they only had 27,506 fans. About how many more fans did they have last week than this week? Estimate to the nearest thousand. 10 points

Estimate of 32,407　　Estimate of 27,506　　Difference of estimates

$$\boxed{32,000} - \boxed{28,000} = \boxed{}$$

⟨**Ans.**⟩ _____

---

2 Yesterday 20,158 people went to the amusement park. Today, there were 28,564. About how many more people came today? Estimate to the nearest ten-thousand. 10 points

Estimate of 28,564　　Estimate of 20,158　　Difference of estimates

$$\boxed{} - \boxed{} = \boxed{}$$

⟨**Ans.**⟩ _____

---

3 There are 126,753 men and 114,397 women in Merriam's town. About how many more men than women are there? Estimate to the nearest ten-thousand. 10 points

⟨**Ans.**⟩ _____

---

4 Susan's father is looking at 2 cars. One costs $37,800, and one costs $56,000. About how much of a difference in price do the cars have? Estimate to the nearest ten-thousand. 10 points

⟨**Ans.**⟩ _____

---

5 3,458 people used Tammy's train stop yesterday. Today, 2,328 people used the stop. About how many more people used the stop yesterday than today? Estimate to the nearest thousand. 10 points

⟨**Ans.**⟩ _____

6 Tim's father is a truck driver. He drove 36,180 miles the year before last, and 42,165 miles last year. What was the approximate difference in the distances he drove the last two years? Estimate to the nearest thousand.

10 points

⟨Ans.⟩ _____

7 The warehouse for Mr. Russell's company held $398,250 worth of books last month. This month, it has $402,500 of books. About how much more value do the books have this month than last month? Estimate to the nearest thousand.

10 points

⟨Ans.⟩ _____

8 In Benji's city, there are 173,500 square meters of park. If there are 268,320 square meters of developed land, about how much more developed land is there than park land? Estimate to the nearest thousand.

10 points

⟨Ans.⟩ _____

9 The widget factory turned out 18,246 widgets yesterday and 21,685 widgets today. About how many more widgets did they produce today? Estimate to the nearest thousand.

10 points

⟨Ans.⟩ _____

10 Sam is training for the 10-kilometer race next month. He ran 25,400 meters last week and 2,860 meters so far this week. About how much more did he run last week? Estimate to the nearest thousand.

10 points

⟨Ans.⟩ _____

Okay, let's switch it up.

Odd & Even

Date   /   /

Name

1   Vicki was running in and out of the classroom door. She crossed the threshold 8 times while she was playing.

5 points per question

(1)   Fill in the appropriate words in the table below.

Number of times across the threshold	0	1	2	3	4
Where Vicki is	room	hall			

(2)   Whenever Vicki was in the classroom, was the number of times across the threshold even or odd?

⟨**Ans.**⟩ _____

(3)   Whenever Vicki was in the hall, was the number of times across the threshold even or odd?

⟨**Ans.**⟩ _____

(4)   Once Vicki had crossed the threshold 8 times, where was she?

⟨**Ans.**⟩ _____

2   For a class exercise, the teacher asked the students to line up boy-girl-boy-girl.

5 points per question

(1)   Are the boys the odd or even numbers in line from the left?

⟨**Ans.**⟩ _____

(2)   Are the girls the odd or even numbers in line from the left?

⟨**Ans.**⟩ _____

(3)   Is the tenth person from the left a boy or a girl?

⟨**Ans.**⟩ _____

3 Mr. Lee decorated his back yard with some black and white stones. He alternated them starting with black, as shown below. Is the fifteenth stone from the left black or white?   10 points

● ○ ● ○ ● ○ ● ○ • • •

⟨Ans.⟩ _____

4 Sheila made a paper necklace by alternating red and yellow paper rings. If she started with red, is the fiftieth ring red or yellow?   10 points

⟨Ans.⟩ _____

5 Rick's father put up the holiday lights, which alternated white and blue bulbs. If the first bulb was white, what color was the thirty-second from the beginning?   15 points

⟨Ans.⟩ _____

6 Our physical education teacher made us stand up so that the boys and girls alternated. The first person in line was a boy.   15 points per question

(1) Was the fifteenth person from the front a boy or a girl?

⟨Ans.⟩ _____

(2) Was the twenty-fourth person from the front a boy or a girl?

⟨Ans.⟩ _____

All I know is that I am odd.
Okay, let's start something new!

Fractions

6

Date   /   /   Name

Level ☆☆   Score   /100

1 In the back of the garage, there is an oil can with $\frac{1}{4}$ liter of oil in it. If you add $\frac{2}{4}$ liter of oil to the can, how much oil will be in the can?   10 points

$$\frac{1}{4} + \frac{2}{4} =$$

⟨**Ans.**⟩ _____

2 Allison's class planted some flowers behind the school. If they planted $\frac{3}{6}$ square foot of tulips and $\frac{2}{6}$ square foot of roses, how many square feet of flowers did they plant in all?   10 points

⟨**Ans.**⟩ _____

3 Mr. Hampton and his son painted the garage. Mr. Hampton painted $\frac{2}{5}$ of the total and his son painted $\frac{1}{5}$ of the total. What fraction of the garage did they finish painting today?   10 points

⟨**Ans.**⟩ _____

4 Cathy is sewing a dress. She used $\frac{3}{7}$ meter of ribbon, and there is $\frac{2}{7}$ meter of ribbon left. How long was the ribbon at first?   10 points

⟨**Ans.**⟩ _____

5 Lisa was shopping for her party, and she put $\frac{3}{4}$ pound of oranges in her basket, which weighed $\frac{1}{4}$ pound. If there is nothing else in her basket, how much does the basket weigh in all?   10 points

⟨**Ans.**⟩ _____

6 The cafeteria used $\frac{5}{7}$ liter of oil this week and they have $\frac{3}{7}$ liter left. How much oil did they have at the beginning of the week?

10 points

⟨Ans.⟩ _____

7 In our pantry, we have $\frac{5}{8}$ pound of rice in a dispenser and $\frac{6}{8}$ pound in a bag. How much rice do we have in all?

10 points

⟨Ans.⟩ _____

8 From Sam's house to his school, it is $\frac{4}{9}$ kilometer, and from the school to the station, it is $\frac{7}{9}$ kilometer. How far is it from Sam's house to the station if he goes past the school?

10 points

⟨Ans.⟩ _____

9 This morning before school, Ramon drank $\frac{2}{8}$ liter of milk, and his sister drank $\frac{1}{8}$ liter. How much did they drink in all?

10 points

⟨Ans.⟩ _____

10 In the crafts classroom, we have $\frac{3}{4}$ yard of white tape and $\frac{2}{4}$ yard of red tape. How much tape is there in the crafts room?

10 points

⟨Ans.⟩ _____

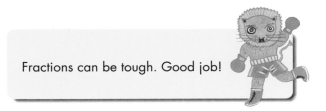

Fractions can be tough. Good job!

1 The chef made a large marinade with $\frac{2}{5}$ liter of soy sauce. If there was $\frac{3}{5}$ liter of soy sauce in his bottle when he started, how much soy sauce does he have left in his bottle?   10 points

$$\frac{3}{5} - \frac{2}{5} =$$

⟨Ans.⟩ _____

2 Lenny's fridge has $\frac{3}{4}$ liter of milk and $\frac{1}{4}$ liter of orange juice in it. What is the difference in the amount of milk and juice in his fridge?   10 points

⟨Ans.⟩ _____

3 Today in gym class, Gina ran $\frac{4}{5}$ mile while Timmy ran $\frac{1}{5}$ mile. What was the difference in the distances they ran?   10 points

⟨Ans.⟩ _____

4 Greg is trying to drink more water to stay hydrated. He drank $\frac{2}{7}$ liter of water yesterday, and $\frac{3}{7}$ liter today. Which day did he drink more, and how much more did he drink that day?   10 points

⟨Ans.⟩ _____

5 City hall is $\frac{8}{9}$ mile from Juliette's house. The train station is $\frac{5}{9}$ mile away. Which is further away, and how much further away?   10 points

⟨Ans.⟩ _____

6 Brenda's little sister was running around the house with the apple juice when she slipped and fell. There used to be $\frac{6}{7}$ liter of juice, but she spilled $\frac{2}{7}$ liter. How much juice is left?

10 points

⟨**Ans.**⟩ _____

7 Mrs. Joseph is having a dinner party today and she used $\frac{3}{4}$ pound of rice. If she had 1 pound of rice to begin, how much rice does she have left?

10 points

⟨**Ans.**⟩ _____

8 Peter's car is running really low on oil. It had $\frac{7}{8}$ liter of oil at the beginning of the month, and now it has $\frac{3}{8}$ liter of oil. How much oil has Peter's car used this month?

10 points

⟨**Ans.**⟩ _____

9 Gina is hiking a 1-kilometer path today with her family. If they have already walked $\frac{3}{5}$ kilometer, how much further do they have to walk?

10 points

⟨**Ans.**⟩ _____

10 Dennis is painting all of his old action figures for fun. He painted $\frac{5}{7}$ of the action figures and then his little brother joined in and did $\frac{2}{7}$ of them. What is the difference between the amount Dennis painted and the amount his brother painted?

10 points

⟨**Ans.**⟩ _____

Take a break if you need one!

Fractions

Date   /   /

Name

---

1 Terri has 2 meters of ribbon. If she divides it into 3 equal pieces, how long will each piece of ribbon be?

10 points

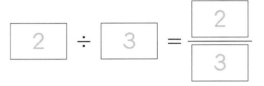 $\boxed{2} \div \boxed{3} = \dfrac{2}{3}$

⟨Ans.⟩  m

---

2 If Terry had 3 meters of ribbon and divided into 4 equal pieces, how long would each piece of ribbon be?

10 points

 $\boxed{} \div \boxed{} = \dfrac{\boxed{}}{\boxed{}}$

⟨Ans.⟩ _____

---

3 Mrs. Ramirez bought 5 pounds of sugar at the store because of a special, but none of her 3 children can carry it alone. How much would each bag weigh if she splits the sugar into 3 bags for them to carry?

10 points

⟨Ans.⟩ _____

---

4 The 7 boys in the class wanted orange juice, so they split the 3 liters of orange juice evenly. The 7 girls wanted apple juice, so they split 4 liters of apple juice evenly. Did the boys or the girls get more juice per person? How much more?

10 points

⟨Ans.⟩ _____

---

5 Later on, the 3 boys split 3 feet of blue gum tape evenly, while 5 girls split 4 feet of red gum tape evenly. Did the boys or the girls get more gum tape? How much more?

10 points

⟨Ans.⟩ _____

---

6 If there are 3 bananas and 5 kiwis in the fruit bowl, how many times more bananas than kiwis are there?   10 points

Bananas   Kiwis

$$\boxed{} \div \boxed{} = \frac{\boxed{}}{\boxed{}}$$

〈Ans.〉 _____

7 There are 7 ducks and 6 geese in the pond. How many more times geese than ducks are there?

10 points

〈Ans.〉 _____

8 Mrs. Johnson seeded 9 square meters of fields, and Mr. Johnson seeded 8 square meters of fields. How many times more area did Mr. Johnson seed than Mrs. Johnson?   10 points

〈Ans.〉 _____

9 Adrian has a green snake and a brown snake. If the green snake is 8 feet long and the brown snake is 3 feet long, how many times longer than the green snake is the brown snake?   10 points

〈Ans.〉 _____

10 Mr. Chiu has a flowerbed in his backyard. If it is 8 feet long and 15 feet wide, how many more times long is it than wide?   10 points

〈Ans.〉 _____

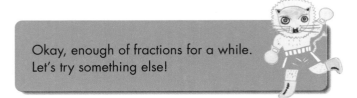

Okay, enough of fractions for a while. Let's try something else!

Decimals

9

Date / /   Name

Level ★★

Score /100

---

1   Max has 7 bottles of soda in his fridge. If each bottle has 0.2 liter in it, how much soda does he have?   10 points

Volume per bottle		Number of bottles		Total volume
0.2	×	7	=	

〈Ans.〉 _____

2   Brook likes licorice a lot. He has 3 strings of licorice today, and each is 0.4 meter long. How much total licorice does he have?   10 points

Length per string		Number of strings		Total length
0.4	×		=	

〈Ans.〉 _____

3   The path around the pond at the park is 0.8 kilometer long. If Diego ran around the pond 6 times, how far did he run?   10 points

〈Ans.〉 _____

4   Mrs. Barnes is heading home from the grocery store with 5 bags. If each bag weighs 0.5 pound, how much weight is she carrying?   10 points

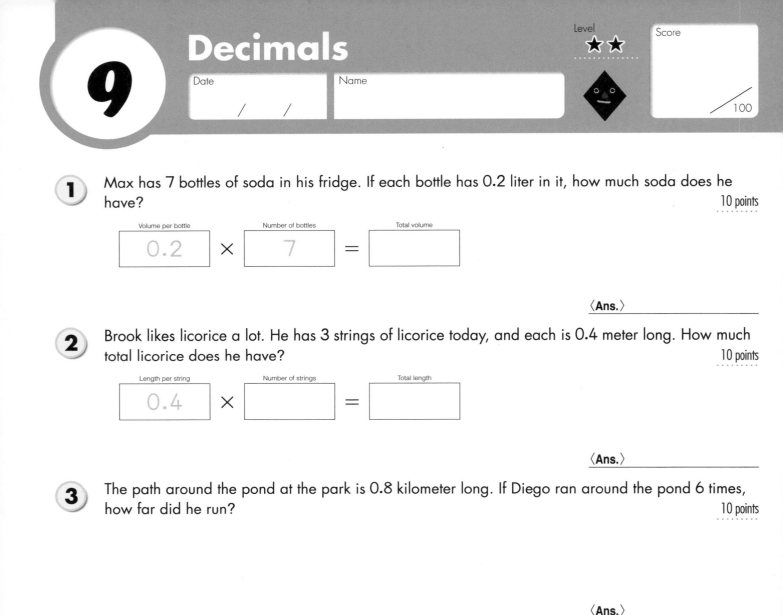

〈Ans.〉 _____

5   The Schroder girls are going to the parade today. Their mother bought a long ribbon and they cut it into 6 equal pieces for their hair. If each ribbon piece was 0.4 foot long, how long was the original ribbon?   10 points

〈Ans.〉 _____

6 Adam is tying up the newspapers for recycling. He used 4 pieces of string that were each 3.2 meters long. How much string did he use?

10 points

⟨**Ans.**⟩ _____

7 Theo biked to the library and back in order to pick up a book he needed for his report. If his house is 4.5 miles away from the library, how far did he ride?

10 points

⟨**Ans.**⟩ _____

8 Mr. Gilchrist is having his family over for the holidays. Since he has a big family, he bought 5 chickens that were 2.8 pounds each. How many pounds of chicken did he buy?

10 points

⟨**Ans.**⟩ _____

9 Clark is trying to lose some weight. He's running 1.6 kilometers every morning before breakfast. If he runs every morning for 25 straight days, how far will he have run in all?

10 points

⟨**Ans.**⟩ _____

10 Jenny is covering seats for her sister's wedding. She needs 18 seat covers that are 2.5 feet long. How much fabric does she need in all?

10 points

⟨**Ans.**⟩ _____

How's it going? Good?

1 Before we went on the hike, our scoutmaster Mr. Harry told us that we only wanted to carry just as much water as we needed. If I split my 1.2 liters of water with 2 other scouts, how much water did each of us get?   10 points

Total volume   Number of people   Volume per person

| 1.2 | ÷ | 3 | = | |

⟨Ans.⟩ _____

2 Greg is working for the cable company and needs to divide his long cable into 5 shorter ones for his customers today. If he has 8.5 meters of cable and he divides it evenly into 5 parts, how long will each piece of cable be?   10 points

Total length   Number of pieces   Length of each piece

| | ÷ | | = | |

⟨Ans.⟩ _____

3 Saskia is helping her father with his garden. He bought 6.4 pounds of sod and divided it up into 4 bags so that she could help him carry the dirt. How much did each bag weigh?   10 points

⟨Ans.⟩ _____

4 Jack has 6 packages to tape up in the mailroom. If he has 7.8 meters of tape for the whole job, how much tape can he use for each package?   10 points

⟨Ans.⟩ _____

5 Mrs. Patterson is planning a square herb garden. She wants the perimeter to be 9.2 feet. How long can each side be?   10 points

Perimeter   Number of sides   Length of a side

| | ÷ | | = | |

⟨Ans.⟩ _____

6 Jenny made 5.4 liters of lemonade for her lemonade stand today, but wants to leave a bottle or two in the fridge inside because it is hot outside. If she divides up all the lemonade into 3 bottles equally, how much lemonade will be in each bottle?

10 points

⟨**Ans.**⟩ _____

7 Sam had 5.4 feet of red vine candy left. If he divided it equally among 6 friends, how long was each person's piece of candy?

10 points

⟨**Ans.**⟩ _____

8 If 4 meters of iron bar weigh 10.4 kilograms, how much does 1 meter of iron bar weigh?

10 points

⟨**Ans.**⟩ _____

9 Ricardo ran around the park 5 times. If he ran 7.5 kilometers in all, what was the circumference of the park?

10 points

⟨**Ans.**⟩ _____

10 Mrs. Randolph has a legendary apple cider. She bought 7 bags of apples and the total weight was 15.4 pounds. How much did each bag weigh?

10 points

⟨**Ans.**⟩ _____

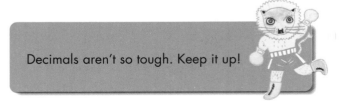

Decimals aren't so tough. Keep it up!

Decimals

11

Date　/　/

Name

Level ★★

Score

/100

1 How many ribbon flowers can the Graham sisters make if they have 8.5 feet of ribbon and each ribbon flower takes 3 feet of tape? How much tape will they have left?　10 points

$$\boxed{8.5} \div \boxed{3} = \boxed{} \quad R \boxed{}$$

⟨**Ans.**⟩ □ flowers, □ ft. remain

2 The pipes burst in Mr. Yang's house. He needs 2 feet of tape to fix a pipe. If he has 13.5 feet of tape, how many pipes can he fix, and how much tape will he have left?　10 points

⟨**Ans.**⟩ _____

3 Hanna's mother is making chocolate cakes for her birthday. If she has 7.5 pounds of chocolate, and it takes 2 pounds per cake, how many cakes can she make, and how much chocolate will remain?　10 points

⟨**Ans.**⟩ _____

4 Tammy wants to make some necklaces with the wire her mother gave her. She has 7.8 feet of wire, and she needs 3 feet for each necklace. How many necklaces can she make, and how much wire will she have left for earrings?　10 points

⟨**Ans.**⟩ _____

5 Omar's cats love to play with string. He has 9.5 feet of string. If he cuts out 4-foot pieces of string, how many pieces will have, and how much will he have left over?　10 points

⟨**Ans.**⟩ _____

6 The cafeteria has a big vat of olive oil in the storage room with 5.4 liters of oil in it. They usually fill a 3-liter bottle with the oil to bring it into the kitchen. How many bottles can they make with the oil in the vat? How much will be left?   10 points

〈Ans.〉 _____

7 Mr. Neese is putting a wire fence around his garden to keep the deer from eating his plants. He has a piece of wire that is 9.4 meters long. If he cuts out a 4-meter piece at a time, how many pieces will he get, and how much will he have left over?   10 points

〈Ans.〉 _____

8 Mrs. Hernandez is excited for her daughter's wedding and wants to start cooking right away. If she went out and bought 18.2 kilograms of rice, and put 5 kilograms in each bag, how many 5-kilogram bags of rice did she get? How much rice was left over?   10 points

〈Ans.〉 _____

9 It is winter, and Lana likes to knit. She bought 24.4 meters of yarn, and then decided it was too much. If she gives each of her friends 7 meters, how many friends will get yarn, and how much yarn will be left?   10 points

〈Ans.〉 _____

10 Hank picked all the apples from his tree and got 19.5 kilograms. If he put 3 kilograms each in a basket, how many baskets can he make, and how many apples will he have left?   10 points

〈Ans.〉 _____

You're doing really well!

23

Decimals

12

Level ★★

Date / /

Name

Score /100

1 Mrs. Chen is cooking eel. She has an eel that is 3.4 feet long. If she divides it into 5 pieces, how long will each piece be? 10 points

$$\boxed{3.4} \div \boxed{5} = \boxed{0.68}$$

⟨Ans.⟩ _____

2 Jim is building a doghouse and has a piece of wood that is 7.4 feet long. If he divides it into 5 equal pieces, how long will each piece of wood be? 10 points

⟨Ans.⟩ _____

3 The Vincents have too much juice to fit in their fridge. If they have 2.6 liters of juice, and divide it into 4 containers equally, how much will be in each container? 10 points

⟨Ans.⟩ _____

4 At the party, we had 1.5 liters of milk. If we split it up among the 6 people left at the party, how much milk did each person get? 10 points

⟨Ans.⟩ _____

5 Greta works with the electric company. Today she has 6.6 feet of cord that she has to split into 4 equal parts. How long will each piece be? 10 points

⟨Ans.⟩ _____

6 The Reyes family is moving. If they have 9 kilograms of plates, and they divide them up into 6 boxes, how much will each box weigh?   10 points

〈**Ans.**〉 _____

7 If you can paint 15 square meters of a wall with 4 liters of paint, how much can you paint with 1 liter of paint?   10 points

〈**Ans.**〉 _____

8 Jed was biking for some exercise, and he circled his block 5 times. If he went 7 kilometers, what is the circumference of his block?   10 points

〈**Ans.**〉 _____

9 While we were driving back from our weekend skiing in the mountains, we passed a farmer selling grapes. We bought 6 kilograms of grapes because we liked them so much. If we split them into 4 packages equally, how much would the grapes in each package weigh?   10 points

〈**Ans.**〉 _____

10 Chef Robinson bought 21 kilograms of potatoes for the Friday night rush at his restaurant. If he put it into 6 bags, how much would the potatoes in each bag weigh?   10 points

〈**Ans.**〉 _____

Now that you understand this, we'll throw rounding in there to make it more complicated! Good luck!

Decimals

Date   /   /

Name

Level ★★

Score
/100

1 It snowed last night, and Mr. Harper needs to get his trucks on the road. He has 8.3 pounds of salt. If he divides the salt equally among his 6 drivers, how much salt will each person get? Round the answer to the nearest tenth.   10 points

$$8.3 \div 6 = 1.\overset{4}{3}8$$

〈**Ans.**〉 _____

2 Mr. Harper also wants to give his drivers plenty of hot coffee before they get out on the road. If he divides his 2.5 liters of coffee equally among his 6 drivers, how much will they each get? Round the answer to the nearest tenth.   10 points

$$2.5 \div 6 = 0.41$$

〈**Ans.**〉 _____

3 3 of the trucks need some fixing before they can get out on the road, too. If Mr. Harper divides the 16.3 feet of tape among the 3 trucks, how long will each piece of tape be? Round the answer to the nearest tenth.   10 points

〈**Ans.**〉 _____

4 The tape didn't fix the 3 trucks, so Mr. Harper went to find some wire. If he only has 7.6 feet of wire, how much wire will each truck get? Round to the nearest tenth.   10 points

〈**Ans.**〉 _____

5 Before he lets the trucks go, Mr. Harper checks the oil in each truck. He has 9 liters of oil that weigh 8.2 pounds. How much does 1 liter of his oil weigh? Round to the nearest tenth.   10 points

〈**Ans.**〉 _____

6 If you can paint 5.4 square meters of the side of the barn with 8 liters of paint, how much barn can you paint with 1 liter of paint? Round to the nearest tenth.

10 points

⟨Ans.⟩ _____

7 Richard hiked 6.5 kilometers even though it was raining lightly. If it took him 3 hours, how many kilometers was he hiking per hour? Round to the nearest tenth.

10 points

⟨Ans.⟩ _____

8 Mikaila can hop 9.5 meters in 4 hops. If she hopped equally far each time, how far did she go in 1 hop? Round to the nearest tenth.

10 points

⟨Ans.⟩ _____

9 Sometimes Billy uses a couple of old iron bars in the bank of his garage to work out. If his 3-meter iron bar weighs 4.4 kilograms, how much does his 1-meter iron bar weigh? Round to the nearest tenth.

10 points

⟨Ans.⟩ _____

10 Julian, the assistant coach, was cleaning up the footballs after practice. He noticed that 8 footballs weighed 1.7 kilograms. How much did each football weigh? Round to the nearest tenth.

10 points

⟨Ans.⟩ _____

Well done! Take a break if you need one.

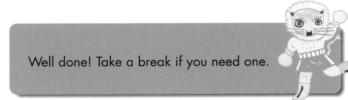

Decimals

14

Date / /

Name

Level ★★

Score
/100

1 Steve has a walking stick that is 45 inches long and a bat that is 30 inches long. How many times longer is his walking stick than his bat? 10 points

Walking stick		Bat		Number of times longer
45	÷	30	=	1.5

⟨Ans.⟩

2 Steve's sister, Wendy, has a walking stick that is 30 inches long and a baton that is 15 inches long. How many more times longer is her baton than her walking stick? 10 points

Baton		Walking stick		Number of times longer
	÷		=	

⟨Ans.⟩

3 Brian's baby brother weighs 35 pounds, while Brian's little brother weighs 56 pounds. How many times heavier is Brian's little brother than his baby brother? 10 points

⟨Ans.⟩

4 Brian's baby brother weighs 35 pounds, while Brian's little brother weighs 56 pounds. How many times heaver is Brian's baby brother than his little brother? 10 points

⟨Ans.⟩

5 Today after class, we helped clean up the park. Adam picked up 15 pounds of trash while Tina picked up 4 pounds. How many times more trash did Adam pick up than Tina? 10 points

⟨Ans.⟩

6 Jamal moved to New York and rented a small room in an apartment. If his room is 2.5 meters long and 5 meters wide, how many more times long than wide is it?   10 points

⟨Ans.⟩ _____

7 After he saved up some money, Jamal found a room that was a little bit bigger. This new room was 9 meters long and 5.4 meters wide. How many more times wide than long was this new room?   10 points

⟨Ans.⟩ _____

8 Andray and Hiro ran around the park until they couldn't any more. Andray managed 7 kilometers, and Hiro ran for 5 kilometers. How many times further than Hiro did Andray run?   10 points

⟨Ans.⟩ _____

9 John is cooking for his family for the holidays. He bought a turkey that weighed 5.4 kilograms and a chicken that weighed 3 kilograms. How many times bigger was the turkey?   10 points

⟨Ans.⟩ _____

10 Ron's family has 1.7 kilograms of salt and 2 kilograms of sugar in the pantry. The amount of salt is how many more times the amount of sugar?   10 points

⟨Ans.⟩ _____

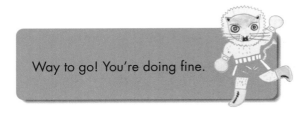

Way to go! You're doing fine.

1 The electrician has 3.6 feet of wire in his truck. If 1 foot of wire weighs 250 ounces, how much does the pile of wire in his truck weigh?   10 points

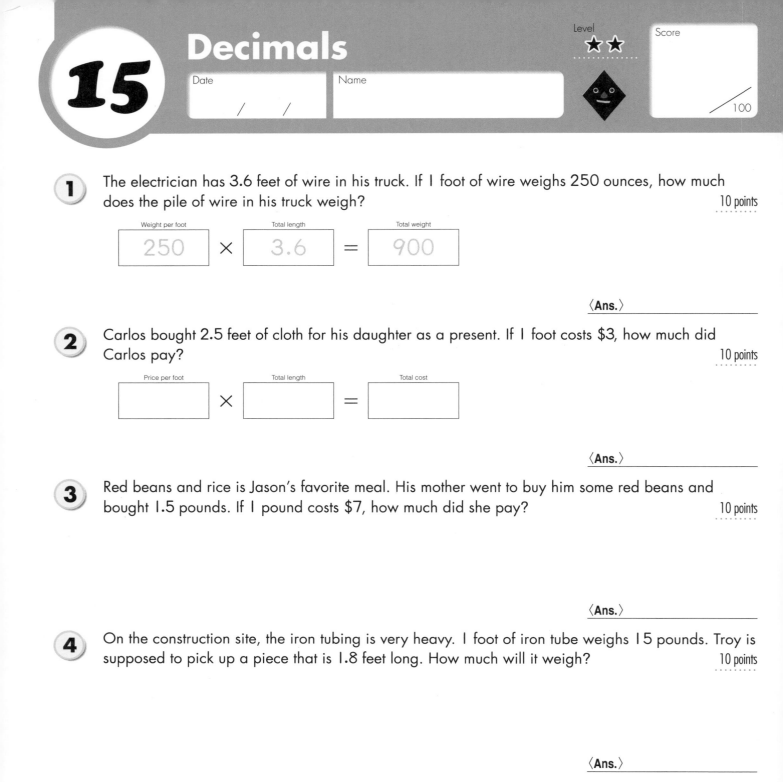

Weight per foot		Total length		Total weight
250	×	3.6	=	900

⟨Ans.⟩ _____

2 Carlos bought 2.5 feet of cloth for his daughter as a present. If 1 foot costs $3, how much did Carlos pay?   10 points

Price per foot		Total length		Total cost
	×		=	

⟨Ans.⟩ _____

3 Red beans and rice is Jason's favorite meal. His mother went to buy him some red beans and bought 1.5 pounds. If 1 pound costs $7, how much did she pay?   10 points

⟨Ans.⟩ _____

4 On the construction site, the iron tubing is very heavy. 1 foot of iron tube weighs 15 pounds. Troy is supposed to pick up a piece that is 1.8 feet long. How much will it weigh?   10 points

⟨Ans.⟩ _____

5 Olive oil at the corner store costs $8 per liter, but Vince only wants 0.6 liter. How much will he have to pay?   10 points

$8 \times 0.6 =$

⟨Ans.⟩ _____

6 Rainer is a beekeeper. 1 liter of his bees' honey weighs 1.2 kilograms. This summer, Rainer got 2.5 liters of honey from his bees. How much did it weigh? 10 points

Weight per liter		Total volume		Total weight
1.2	×	2.5	=	3

⟨Ans.⟩ _____

7 My car gets 9.5 kilometers per 1 liter of gas. I only have 6.5 liters of gas left. How far will my car go? 10 points

Kilometers per liter		Total volume		Total distance
	×		=	

⟨Ans.⟩ _____

8 Father can mow 25.4 square meters of lawn in 1 hour. He's been mowing for 1.5 hours – how many square meters has he mowed? 10 points

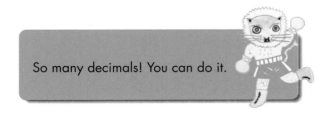

⟨Ans.⟩ _____

9 1 meter of plastic pipe that the plumber is using to fix your bathroom weighs 2.6 kilograms. If he's using 3.4 meters of pipe, how much does it weigh? 10 points

⟨Ans.⟩ _____

10 Jim's family loves soy sauce. 1 liter of it weighs 1.2 kilograms. If Jim's family used 0.8 liter last month, how much did it weigh? 10 points

⟨Ans.⟩ _____

So many decimals! You can do it.

Decimals

Date / / 　　Name

Level ★★

Score /100

1 Laura's little sister weighs 38 pounds, and her brother weighs 1.5 times as much. How much does her brother weigh? 　10 points

Laura's little sister's weight　　How many times bigger　　Her brother's weight

$$38 \times 1.5 = \boxed{}$$

〈Ans.〉_____

2 From Derek's house to the train station, it is 450 meters. If the school is 1.2 times further away, how far is his house from school? 　10 points

Distance to the station　　How many times further　　Distance to school

$$\boxed{} \times \boxed{} = \boxed{}$$

〈Ans.〉_____

3 Manny got 2 packages in the mail on his birthday. The little one weighs 16 pounds, and the big one is 2.5 times as heavy. How much does the big one weigh? 　10 points

〈Ans.〉_____

4 Rich is trying to keep track of his family's use of heating oil. Yesterday, they used 12 liters. Today, they used 1.4 times as much. How much did they use today? 　10 points

〈Ans.〉_____

5 Selena has a big test coming so she studied 3 hours. If her younger brother studied 0.5 times as much as her, how long did he study? 　10 points

〈Ans.〉_____

32　© Kumon Publishing Co., Ltd.

6 Melanie is excited. Her mother promised to make Melanie a dress for her birthday party. They went and bought 1.8 meters of cloth and paid $9. How much did they pay per meter?   10 points

Cost		Length bought		Cost per meter
9	÷	1.8	=	5

⟨Ans.⟩ _____

7 Steven wants to go rock climbing, so he went to buy 2.5 meters of rope. If he paid $10, how much did he pay per meter?   10 points

Cost		Length bought		Cost per meter
	÷		=	

⟨Ans.⟩ _____

8 Now Melanie and her mother are going to buy sugar for the cake. If they bought 2.5 kilograms of sugar for $6, how much did they pay per kilogram?   10 points

⟨Ans.⟩ _____

9 The warehouse has 36 liters of soy sauce. They divide the sauce up into 1.8 liter bottles for all of the restaurants they service. How many bottles do they need for all the soy sauce they have now?   10 points

⟨Ans.⟩ _____

10 Mr. Delgado is leading a science experiment today. 0.9 meter of the copper wire they are using weighs 306 grams. If they have 1 meter for the experiment, how much does it weigh?   10 points

Weight of wire		Length of wire		Weight per meter
306	÷	0.9	=	

⟨Ans.⟩ _____

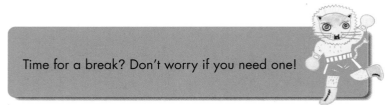

Time for a break? Don't worry if you need one!

Decimals

1 Mr. Fielder is the foreman on a building site. They are building a new school, and right now the plumbing needs to go in. If 1.6 feet of the iron pipe they have weighs 6.4 pounds, how much does 1 foot weigh?   10 points

Total weight		Length of iron pipe		Weight per foot
6.4	÷	1.6	=	

〈Ans.〉

2 Mr. Fielder goes to check on the painters. They have 3.5 liters of paint which weigh 4.2 pounds. How much would 1 liter weigh?   10 points

Total weight		Volume of the paint		Weight per liter
	÷		=	

〈Ans.〉

3 It is time to check on the electricians as they work on the second floor. Mr. Fielder sees that 4.5 feet of their wire weighs 2.7 pounds. How much would 1 foot of their wire weigh?   10 points

〈Ans.〉

4 The heating oil man comes and Mr. Fielder has to negotiate with him over the price. The oil man wants to drop off 3.5 pounds, which is 4.2 liters of oil. How much oil would 1 pound of oil be?   10 points

〈Ans.〉

5 An accident happened in the gym and Mr. Fielder went to investigate. A 1.4-pound aluminum bar fell and broke. If the bar was 2.8 feet long, how much would a 1-foot bar weigh?   10 points

〈Ans.〉

6 Mary Kate is supposed to pick up the watermelons for the party. Each one weighs 1.2 kilograms. If she bought 28.8 kilograms of watermelons in all, how many watermelons did she buy?   10 points

Total weight		Weight of watermelon		Number of watermelons
28.8	÷	1.2	=	

⟨Ans.⟩ _____

7 The baker went to buy sugar from the warehouse store. Each bag was 1.5 pounds, and he bought 37.5 pounds of sugar in all. How many bags did he guy?   10 points

Total weight		Weight of a bag		Number of bags
	÷		=	

⟨Ans.⟩ _____

8 Ted is training for a triathlon and bikes 2.5 kilometers every day. So far he's already biked 77.5 kilometers. How many days has he been training?   10 points

⟨Ans.⟩ _____

9 The art teacher has 13.4 meters of tape, and decided to divide it into smaller pieces that were 2.5 meters each. How many smaller pieces did he make, and how much tape was remaining?   10 points

- - - - - 13.4 m - - - - -
- 2.5 m -   remaining

13.4	÷	2.5	=	5	R	0.9

⟨Ans.⟩ _____

10 The milkman has 21.9 liters of milk to deliver. If he divides it into 1.8-liter bottles, how many bottles will he make? How much milk will be left over?   10 points

	÷		=		R	

⟨Ans.⟩ _____

Just a little bit more work with decimals. You can do it!

Decimals

1   Dan is planning to make a pen for his chickens in the back yard. If the pen will be 5.8 feet long and 8.7 feet wide, how many times more wide than long will the pen be?   10 points

Width   Length   How many times wider

$$8.7 \div 5.8 = \boxed{}$$

⟨Ans.⟩ _____

2   Beetles are very strong bugs that are capable of pulling objects that are many times larger than they are. A 6.5-ounce beetle can easily carry a 104-ounce piece of food. How many times larger than the beetle is the piece of food?   10 points

Weight of food   Weight of beetle   How many times larger

$$\boxed{} \div \boxed{} = \boxed{}$$

⟨Ans.⟩ _____

3   Mrs. Hughes needs a lot of flour, but can't decide between buying the big box or many smaller boxes. The big box of flour weighs 28.5 pounds, and the little box weighs 1.5 pounds. How many times more does the big box weigh?   10 points

⟨Ans.⟩ _____

4   Jason and John are having fun with a tape measure. Jason's driveway is 8.6 feet long. John's driveway is 12.9 feet long. How many times longer is John's driveway?   10 points

⟨Ans.⟩ _____

5   The contractor that the Hampton family called in is running out of tape. He has 8.6 feet of electrical tape and 4.3 feet of duct tape. How many more times longer is his duct tape than his electrical tape?   10 points

⟨Ans.⟩ _____

6 Ava's father weighs 54.6 kilograms, which is 1.5 times Ava's weight. How much does Ava weigh?

10 points

Father's weight		How many times heavier		Ava's weight
54.6	÷	1.5	=	

⟨**Ans.**⟩ _____

7 Justin is training for the long jump and can jump 4.5 meters. His little brother is jealous, so he jumps sometimes after his brother. If Justin can jump 1.8 times further than his little brother, how far can his little brother jump?

10 points

Length of Justin's jump		How many times further		Length of Justin's brother's jump
	÷		=	

⟨**Ans.**⟩ _____

8 Ashok's father is 168.6 centimeters tall, and that is 1.2 times taller than Ashok. How tall is Ashok?

10 points

⟨**Ans.**⟩ _____

9 The biggest pumpkin Farmer Williams has ever seen weighed 28.5 kilograms. That is 1.5 times the size of the biggest pumpkin he ever grew himself. How much did the biggest pumpkin Farmer Williams ever grew weigh?

10 points

⟨**Ans.**⟩ _____

10 Mrs. Pearman is looking at heating oil tanks for her family's new house. The big oil tank holds 5.4 liters, which is 1.2 times bigger than the small tank. How much oil does the smaller tank hold?

10 points

⟨**Ans.**⟩ _____

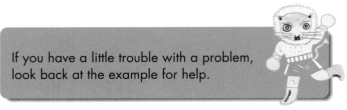

If you have a little trouble with a problem, look back at the example for help.

19 Decimals

Date / /

Name

Level ★★

Score ___/100

1 Irene is a baker and needs to know how much her ingredients weigh by volume. She knows that 10 liters of flour weighs 8 pounds. If she has 1.8 liters of flour, how much does it weigh? 5 points per question

(1) How much does 1 liter of flour weigh?

〈Ans.〉 _____

(2) How much does 1.8 liters of flour weigh?

〈Ans.〉 _____

2 Andy works with computer cords all day. He knows 30 inches of cord weigh 21 ounces. If he has 70 inches of cord today, how much does it weigh? 10 points

$21 \div 30 \times 70 =$

〈Ans.〉 _____

3 If there is 45.3 ounces of salt in every 1.5 pounds of sea water, how much salt is in 5.5 pounds of sea water? 10 points

〈Ans.〉 _____

4 Since Holly hikes long distances, she has to be careful about much weight she puts in her backpack. She knows that 12.4 feet of rope weighs 0.8 pound. If she has 22.6 pounds of rope, how long is it? 10 points

〈Ans.〉 _____

5 Rick can paint 0.6 square foot of wall with 1.2 liters of paint. If he wants to paint 2.5 square feet of wall, how much paint does he need to buy? 10 points

〈Ans.〉 _____

6 Mrs. Bush is packing up her books. She put a bunch of dictionaries in a 1.4-kilogram box, and now the box weighs 18.2 kilograms. If one dictionary weighs 1.2 kilograms, how many dictionaries did she put in the box? 10 points

$$(18.2 - 1.4) \div 1.2 =$$

⟨Ans.⟩ _____

7 Frank has to deliver juice today. He put a bunch of bottles in a 1.1 kilogram box, and then forgot how many bottles he put in. If the total weight of the box is 20.9 kilograms, and one bottle weighs 1.8 kilograms, how many bottles are in the box? 10 points

⟨Ans.⟩ _____

8 Carlos is a chef at a local restaurant and has to buy lots of beans every week. This week, he put 12.5 liters of beans into a bag, and the total weight was 10 kilograms. If the weight of the bag was 0.2 kilogram, how much was the weight of 1 liter of beans? 10 points

⟨Ans.⟩ _____

9 The sink downstairs is very strange. The hot water runs from a narrow faucet at 10.6 liters per minute, while the cold water runs from a thick faucet at 15.8 liters per minute. If you ran both faucets for 5.5 minutes, how much water would you use? 10 points

⟨Ans.⟩ _____

10 My car goes 8.6 kilometers with 1 liter of gas. When I went to fill up today, I had 12.5 liters of gas and then I added 20.8 liters. If I used up all the gas in my car, how far could I go? 10 points

⟨Ans.⟩ _____

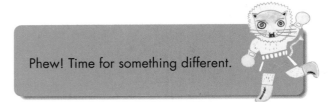

Phew! Time for something different.

Ratios

Date   /   /

Name

Level ★★

Score
/100

Don't forget!

4 boys and 6 girls

The ratio of boys to the total number of children in the picture on the left is 0.4, meaning that the number of boys is 0.4 times the total.

The ratio is a comparison between two similar quantities, A and B. A (the quantity to be compared) ÷ B (the base quantity) = Ratio

⟨Example⟩ 4 ÷ 10 = 0.4

1 What is the ratio of boys to each total below? What is the ratio of girls to each total below?

8 points per question

(1)

7 boys and 3 girls

The ratio of boys to the total (7 ÷ 10 = ☐)

The ratio of girls to the total (3 ÷ 10 = ☐)

(2)

3 boys and 2 girls

The ratio of boys to the total (3 ÷ 5 =)

The ratio of girls to the total (2 ÷ 5 =)

(3)

1 boy and 4 girls

The ratio of boys to the total (   )

The ratio of girls to the total (   )

(4)

1 boy and 3 girls

The ratio of boys to the total (   )

The ratio of girls to the total (   )

(5)

6 boys and 9 girls

The ratio of boys to the total (   )

The ratio of girls to the total (   )

2 Rewrite each ratio below as a percentage.   3 points per question

(1) 0.04   (4%)   (2) 0.09   (%)

(3) 0.16   (   )   (4) 0.45   (   )

(5) 0.5   (   )   (6) 1   (   )

(7) 1.09   (   )   (8) 2   (   )

(9) 0.006   (   )   (10) 0.807   (   )

3 Rewrite each percentage below as a ratio.   3 points per question

(1) 3%   (0.03)   (2) 9%   (   )

(3) 12%   (   )   (4) 60%   (   )

(5) 97%   (   )   (6) 100%   (   )

(7) 250%   (   )   (8) 307%   (   )

(9) 0.8%   (   )   (10) 60.2%   (   )

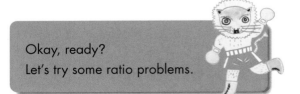

Okay, ready?
Let's try some ratio problems.

1　Mr. Morris wants to join a supper club that allows only a limited number of people in every year. They will take 20 new members this year, and 16 people have applied. How many times more applicants than openings do they have?　10 points

⟨Ans.⟩ _____

2　Our track club can only have 20 members every year. This year, we received 16 applications. What is the ratio of the number of applicants to the total number of members allwed?　10 points

$16 \div 20 = 0.8$

⟨Ans.⟩　　0.8

3　There are 35 students normally in Julia's class. If there are 7 students absent today, what is the ratio of absent students to the whole class?　10 points

$7 \div 35 =$

⟨Ans.⟩ _____

4　This year, Tara planted 40 sunflower seeds, and got 35 sunflowers. What is the ratio of number of sunflowers to the number of seeds she planted?　10 points

⟨Ans.⟩ _____

5　Gina has 36 students in her class and 18 of them have goldfish. What is the ratio of students with goldfish to the whole class?　10 points

⟨Ans.⟩ _____

6 If there is 9 ounces of salt in each 150 ounces of salt water, what is the ratio of the amount of salt to the total amount of salt water?　　10 points

〈Ans.〉 _____

7 Mike's soccer team played 20 games and won 15. What was the ratio of wins to total games played?　　10 points

〈Ans.〉 _____

8 In the 200 yards of fields at my school, there are 40 yards of planted flowers. What is the ratio of planted flowers to the total amount of fields?　　10 points

〈Ans.〉 _____

9 Brenda bought a dress for $500. If the shop discounted the dress $20, what was the ratio of her discount to the original price?　　10 points

〈Ans.〉 _____

10 The tallest building in Tom's town is 64.8 meters tall. Tom's school is 13.5 meters tall. What is the ratio of the height of the tallest building in town to the height of the school?　　10 points

〈Ans.〉 _____

This is not so bad, right? Good!

　　43

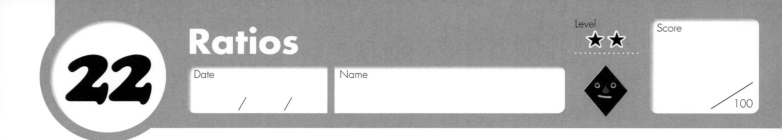

22 Ratios

Level ★★

Score
/100

Date / /

Name

1 Rob weighs 35 pounds. His big brother weighs 1.2 times as much. How much does Rob's brother weigh? 10 points

$35 \times 1.2 =$

⟨Ans.⟩

2 Mr. Thomas is making a bird feeder. On the base of the feeder, the width is 1.4 feet. If the length is 1.5 times the width, how long is the base of the bird feeder? 10 points

⟨Ans.⟩

3 Brenda is at the airport with her family. Her bag weighs 35 pounds. Her sister's bag weighs 0.9 times as much as her bag. How much does her sister's bag weigh? 10 points

$35 \times 0.9 =$

⟨Ans.⟩

4 There are 180 fifth grade students at Terry's school. If the ratio of boys to the total is 0.6, how many boys are there in the fifth grade at his school? 10 points

⟨Ans.⟩

5 We had 48 liters of heating oil at the beginning of the week, but it was a cold week and the ratio of the oil we used to the whole was 0.7. How much oil did we use? 10 points

⟨Ans.⟩

6 Lucy is reading her book about worms today. If the book has 240 pages, and she read 0.3 times the whole today, how many pages did she read?

10 points

⟨**Ans.**⟩ _____

7 Of the 860 students at Rhonda's school, 0.2 times the whole live in nearby Littleton. How many students live in nearby Littleton?

10 points

⟨**Ans.**⟩ _____

8 Doyle is 140 centimeters tall. If the ratio of his brother's height to his is 0.8, how tall is his brother?

10 points

⟨**Ans.**⟩ _____

9 The total number of people in Littleton is 6,855. 1.2 times as many people live in Bigton nearby. How many people live in Bigton?

10 points

⟨**Ans.**⟩ _____

10 My aunt, who has an apple farm, sent us 400 apples. 0.1 times the whole was rotten when the packages arrived. How many apples were rotten?

10 points

⟨**Ans.**⟩ _____

Keep on trucking! You can do it.

1 Kim's father is a fisherman. He caught a bass and a big catfish last weekend. If the bass weighed 49 pounds, and weighed 1.4 times as much as catfish, how much did the catfish weigh? 10 points

$49 \div 1.4 =$

⟨Ans.⟩ _____

2 This weekend, Kim's father caught a pike that weighed more than the 49 pounds that the bass weighed last week. If last week's bass weighed 0.7 times as much as the pike, how much did the pike weigh? 10 points

$49 \div 0.7 =$

⟨Ans.⟩ _____

3 Corey's baseball team won 28 games this season. The ratio of the number of wins to the total games played this season was 0.4. How many games did they play in all? 10 points

⟨Ans.⟩ _____

4 Kim is working on her math homework. She has finished 42 problems, which is 0.6 times the whole. How many problems does she have altogether? 10 points

⟨Ans.⟩ _____

5 Lenny is cleaning out his fish tank. He just put 4.8 liters of water back in the tank, and that is only 0.2 times the whole. How much water can his tank hold? 10 points

⟨Ans.⟩ _____

6 The number of absent students today was 0.2 times the total number of grade 5 students. If 24 people were absent, how many total students are in grade 5?   10 points

〈Ans.〉 _____

7 Flo has been reading her book all vacation. If she has finished 204 pages, and that is 0.8 times the whole, how many pages does her book have in all?   10 points

〈Ans.〉 _____

8 In today's parade, there are 306 boys from different clubs and groups. If they represent 0.4 times the whole number of children in the parade, how many children are in the parade?   10 points

〈Ans.〉 _____

9 Cathy is 136 centimeters tall. She is 0.8 times as tall as her brother. How tall is her brother?   10 points

〈Ans.〉 _____

10 Hunter needs a new computer, and also a set of encyclopedias. If the computer costs $2,520, and is 1.4 times as much as the set of reference books, how much do the encyclopedias cost?   10 points

〈Ans.〉 _____

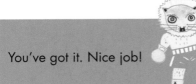

You've got it. Nice job!

Ratios

1 There are 40 students in Serena's class and 16 have dogs at home. What ratio of students has dogs as pets?   10 points

⟨**Ans.**⟩ _____

2 Of the 40 students in Serena's class, 24 have had cavities this year. What percentage of the students in her class has had a cavity this year?   10 points

$24 \div 40 = 0.6$

> Change the ratio into a percentage. 1 is 100%.
> 0.1 is 10%, 0.2 is 20%, ······.
> 0.01 is 1%, 0.02 is 2%, ······.

⟨**Ans.**⟩ 60%

3 There were 60 books in the library in Emily's classroom. They added 18 books this month. What percentage of the old amount of books did they add this month?   10 points

⟨**Ans.**⟩ _____

4 Allison got 45 out of 50 problems correct on her math test. What percentage did she get correct?   10 points

⟨**Ans.**⟩ _____

5 Some bodies of water have more salt than others. The Dead Sea, for example, is known for the saltiness of the water. If a body of water has 30 ounces of salt in 150 ounces of salt water, what percentage of the whole is salt?   10 points

⟨**Ans.**⟩ _____

6 Littleton is a farming town as it had 3,600 square meters of farm last year. This year, the town added 900 square meters of farmland. What percentage increase over last year does this new farmland represent?

10 points

〈Ans.〉 _____

7 Eduardo was looking at some gloves last week that cost $24. When he went in to buy them, he discovered that the price increased $3.6. What percentage did the price increase over the old price?

10 points

〈Ans.〉 _____

8 I bought some canned soup because I was sick. Even though the soup cost $4 a can, the clerk felt bad and discounted each can $0.50. What percentage is the discount of the old price?

10 points

0.125 is 12.5%

〈Ans.〉 _____

9 Parker's town had 6,420 inhabitants last year. If the town now has 321 more people this year, what is the percentage increase over last year?

10 points

〈Ans.〉 _____

10 Over the holidays, the stores in Sheila's town discount many of their items. Sheila was looking at a sweater that was discounted $4. If the original price was $32, what percentage was the discount of the original price?

10 points

〈Ans.〉 _____

Do you understand the relationship between ratio and percentage? Good!

1 Of the 35 students in Mary's class, 20% have broken an arm or a leg before. How many students have broken a limb ?

10 points

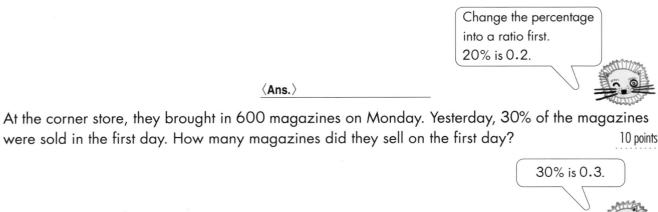

Change the percentage into a ratio first. 20% is 0.2.

⟨Ans.⟩ _____

2 At the corner store, they brought in 600 magazines on Monday. Yesterday, 30% of the magazines were sold in the first day. How many magazines did they sell on the first day?

10 points

30% is 0.3.

⟨Ans.⟩ _____

3 Juliette's classroom has 80 books in the library. If 60% of them are novels, how many novels do they have?

10 points

⟨Ans.⟩ _____

4 The pants I wanted to buy just went up in price. They used to cost $45 and are now going up 20% in price. How much more will the pants cost now?

10 points

⟨Ans.⟩ _____

5 Mrs. Ramirez has a nice garden that is 24 square feet big. If she planted tulips in 40% of her garden, how many square feet of tulips did she plant?

10 points

⟨Ans.⟩ _____

6   Mr. Kwan's train can only fit 120 people. If the train is 70% full, how many people are on the train now?

10 points

⟨**Ans.**⟩ _____

7   There is a 20%-off sale at the toy store today. The old game system package I wanted was originally $200. How much cheaper is it now?

10 points

⟨**Ans.**⟩ _____

8   Chris has a big family. Last year, they ate 150 pounds of potatoes. This year, they ate 120% of that amount. How many potatoes did they eat this year?

10 points

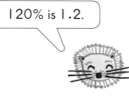

120% is 1.2.

⟨**Ans.**⟩ _____

9   Last year, 640 people took yoga classes at Zack's studio. This year, 5% more people came to take classes from him. How many more students does he have this year?

10 points

5% is 0.05.

⟨**Ans.**⟩ _____

10   My apple is 86% water. If it weighs 180 grams, how much does the water weigh?

10 points

86% is 0.86.

⟨**Ans.**⟩ _____

This is an important real-life skill. Good job!

1 Today, 12 people missed the 8:05 train. If this is 10% of all the people who take that train, how many people normally take the 8:05 train?   10 points

$12 \div 0.1 =$

10% is 0.1.

⟨Ans.⟩ _____

2 21 people in line at the grocery store are in the express lanes. If they represent 60% of all the people in line, how many people are in line at the grocery store in all?   10 points

⟨Ans.⟩ _____

3 Aaron didn't have enough money to completely fill up his car, but he needed some gas. If he bought 6 liters of gas, and that was 20% of a full tank, how much gas can he fit into his gas tank?   10 points

⟨Ans.⟩ _____

4 I have 12 comic books on my bookshelf, and that is 30% of all of my books. How many books do I have in all?   10 points

⟨Ans.⟩ _____

5 Gordon wanted to paint his bike, so he bought a can of $3 paint. If that was 40% of his money, how much money did he have at first?   10 points

⟨Ans.⟩ _____

6 Sue's father has a nice back yard, but he wanted to have a flowerbed near the house. He used 80 square meters, which was 10% of his whole back yard. How big is his back yard?   10 points

⟨**Ans.**⟩ _____

7 240 girls are in Maggie's summer camp. If that is 40% of the total number of children at the camp, how many children are at the camp?   10 points

⟨**Ans.**⟩ _____

8 Mr. Gregory's farm produced 3,600 kilograms of oranges in the month of June. That was 120% of the amount that they produced last year for the same month. How many oranges did they produce last June?   10 points

⟨**Ans.**⟩ _____

9 Rahad is trying to put on a little weight so he can make his wrestling team. This April, he weighs 41.8 kilograms. If that is 110% of last year's weight, how much did he weigh last April?   10 points

⟨**Ans.**⟩ _____

10 The shopkeeper at the corner store had a delivery of tomatoes and 140 were rotten. If this was 10% of all the tomatoes he got, how many tomatoes did he have delivered in all?   10 points

⟨**Ans.**⟩ _____

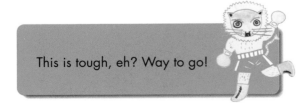

This is tough, eh? Way to go!

Ratios

1 When Nelson's throat is sore, his mother makes him salt water to gargle. This time, she made the salt water with 90 ounces of water and 10 ounces of salt. What percentage of the whole weight is the salt?   10 points

Weight of salt   Weight of salt water   Ratio

$10 \div (90 + 10) = \boxed{}$

⟨Ans.⟩ _____

2 Mrs. Delano made some lemonade with a mix packet. If she put 30 ounces of mix in for 120 ounces of water, what percentage of the whole weight was the mix?   10 points

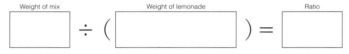

Weight of mix   Weight of lemonade   Ratio

$\boxed{} \div (\boxed{}) = \boxed{}$

⟨Ans.⟩ _____

3 The chef in the restaurant was making a soup. He started with 480 ounces of water and 20 ounces of salt. What percentage of the whole weight was the salt?   10 points

⟨Ans.⟩ _____

4 Mrs. Morris had the class plant sunflower seeds in the spring. If 45 seeds grew, and 5 did not, what percentage of the seeds grew?   10 points

⟨Ans.⟩ _____

5 Diana was at basketball practice and her coach asked her to practice her shot. She made 62 baskets and missed 18 times. What percentage of the time did she make the shot?   10 points

⟨Ans.⟩ _____

6 The music player that Dana wanted for her birthday was $500 last year, but now it is $50 more. What percentage of last year's price is the new price? 10 points

(500 + 50) ÷ 500 =

This year's price ÷ Last year's price = Ratio

1.1 is 110%.

⟨**Ans.**⟩

7 Last year, the jacket I wanted was $300. Now it is $60 higher. What percentage of last year's price is the new price? 10 points

( ) ÷ =

This year's price ÷ Last year's price = Ratio

⟨**Ans.**⟩

8 There were 456 students in Rahim's school last year. If 114 students were added this year, what percentage of last year's total is this year's student body? 10 points

⟨**Ans.**⟩

9 Peggy is excited about a famous violin player that is coming to her college. He was supposed to play a very limited concert for 250 people, but they just opened up the hall and added 150 seats. What percentage of the original number of people will get to see the show now? 10 points

⟨**Ans.**⟩

10 The cafeteria used 2,800 kilograms of potatoes last year. This year, they used 600 kilograms more. What percentage of last year's potatoes did they use this year? Round the answer to the nearest whole number. 10 points

⟨**Ans.**⟩

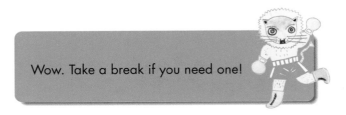

Wow. Take a break if you need one!

1 The dress Theresa wants is on sale for $270 from $300. What percentage of the orginal price is the discount?   10 points

$$\left(\underset{\text{The discount}}{300 - 270}\right) \div \underset{\text{The original price}}{300} = \underset{\text{Ratio}}{\boxed{}}$$

⟨Ans.⟩ _____

2 Webb needs a suit for graduation. He had been looking at a suit for $200, and he bought it for $180. What percentage of the original price was the discount?   10 points

$$\left(\underset{\text{The discount}}{\boxed{}}\right) \div \underset{\text{The original price}}{\boxed{}} = \underset{\text{Ratio}}{\boxed{}}$$

⟨Ans.⟩ _____

3 Mr. Alexander was looking at a nice new television for $500. He saw a sale advertised in the newspaper and got the television for $400. What percentage off did he get?   10 points

⟨Ans.⟩ _____

4 800 people came to see Jimmy's band play. If there were 380 women in the audience, what percentage of the audience were men?   10 points

⟨Ans.⟩ _____

5 In chemistry class today, you made a solution of water and sodium that weighed 250 grams. If you put in 200 grams of water, what percentage of the total weight was the sodium?   10 points

⟨Ans.⟩ _____

6 Tonight, 560 people attended Ron's play. That means 60 more people came tonight than last night. What percentage increase was tonight's attendance over last night's attendance? 10 points

Increased attendance | Yesterday's attendance | Ratio

$$60 \div (560 - 60) = \boxed{}$$

⟨Ans.⟩ _____

7 The price of a plane ticket to visit Tony's grandmother increased $40 to $540 this year. What percentage increase was this new price over last year's price? 10 points

Price increase | Last year's price | Ratio

$$\boxed{} \div (\boxed{}) = \boxed{}$$

⟨Ans.⟩ _____

8 There are 48 seals in the tank at the zoo. That is 8 more seals than last year. What percentage did the seal population increase over last year? 10 points

⟨Ans.⟩ _____

9 Bain's Bike Shop sold 75 bikes this month, 15 more than last month. What percentage did their sales increase over last month? 10 points

⟨Ans.⟩ _____

10 Frey's farm added 300 square meters of potatofields this year, and now he has 4,300 square meters of potatofields. What percentage did the potatofields increase over last year? 10 points

⟨Ans.⟩ _____

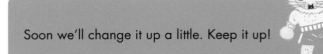

Soon we'll change it up a little. Keep it up!

Ratios

Level
★★

Date
/   /

Name

Score
/100

29

1 The owner of the furniture store bought a couch for $300. She would like to add 20% to the sales price when she puts the couch for sale in her store. What price should put on the couch?   10 points

$$300 \times 1.2 = 360$$

Remember that adding 20% means that the new price will be 120% of the old price.

⟨Ans.⟩ _____

2 At the music store, the owner brought in a guitar that he had paid $400 for. He wants to make 10% from selling the guitar. What price should he put on the guitar?   10 points

⟨Ans.⟩ _____

3 There were 480 students in Alik's school last year. This year, there was a 10% increase over last year's total. How many students are in his school this year?   10 points

⟨Ans.⟩ _____

4 For Richard's art, he needs a scanner. The scanner cost $600 last year, but the price increased 30% over last year's price. How much is the scanner this year?   10 points

⟨Ans.⟩ _____

5 Ken's father stayed at a hotel that cost $130 a night. He has to pay 10% as a service fee. How much is a night at his hotel altogether?   10 points

⟨Ans.⟩ _____

6 Andrea's mother paid $800 for car insurance this year, but then she got a 20% discount because she is a teacher. How much did she end up paying?   10 points

$$800 \times 0.8 = 640$$

Remember, a 20% discount means the new price is 80% of the old price.

⟨Ans.⟩ _____

7 Judy's play had 480 people in the audience yesterday, but a bad review came out today. 20% fewer people came today. How many people came to her play today?   10 points

⟨Ans.⟩ _____

8 Nikki got her shoes with a 25% discount. If the original price was $28, how much did she actually pay?   10 points

⟨Ans.⟩ _____

9 35% of the students at my school have had a cavity in the last year. If there were 600 students at my school, how many of them did have a cavity in the last year?   10 points

⟨Ans.⟩ _____

10 Elle's father tilled 68% of his field today. If he has 600 square meters of fields, how much did he till today?   10 points

⟨Ans.⟩ _____

Okay! Now for something a little different.

Graphs, Percentages & Ratios

Level ★★

Date   /   /

Name

Score
/100

1 The band graph below shows the ratio of each of the different types of vehicles that passed by the school today. 200 vehicles passed the school in all.

5 points per question

Types of vehicles that passed the school today

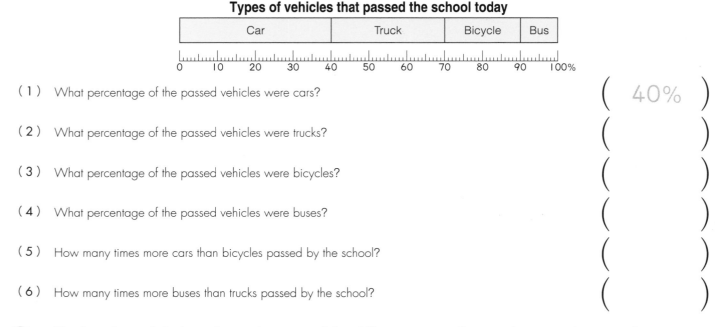

| Car | Truck | Bicycle | Bus |

0 10 20 30 40 50 60 70 80 90 100%

(1) What percentage of the passed vehicles were cars? (40%)

(2) What percentage of the passed vehicles were trucks? (   )

(3) What percentage of the passed vehicles were bicycles? (   )

(4) What percentage of the passed vehicles were buses? (   )

(5) How many times more cars than bicycles passed by the school? (   )

(6) How many times more buses than trucks passed by the school? (   )

2 The band graph below shows the ratio of the different types of crops that were harvested in a certain county.

5 points per question

Types of crops harvested

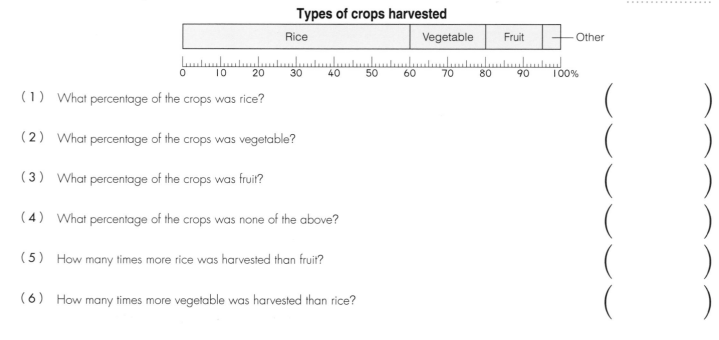

| Rice | Vegetable | Fruit |—Other

0 10 20 30 40 50 60 70 80 90 100%

(1) What percentage of the crops was rice? (   )

(2) What percentage of the crops was vegetable? (   )

(3) What percentage of the crops was fruit? (   )

(4) What percentage of the crops was none of the above? (   )

(5) How many times more rice was harvested than fruit? (   )

(6) How many times more vegetable was harvested than rice? (   )

3 The band graph below shows the ratio of expenditures in the Norton family per month.

4 points per question

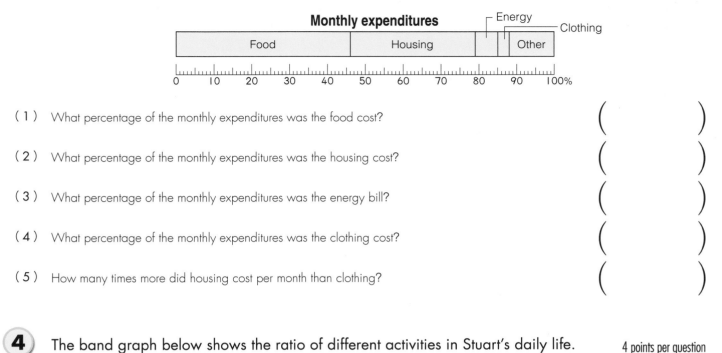

Monthly expenditures

Food	Housing		Other

Energy ⌐

Clothing

0 10 20 30 40 50 60 70 80 90 100%

(1) What percentage of the monthly expenditures was the food cost? (   )

(2) What percentage of the monthly expenditures was the housing cost? (   )

(3) What percentage of the monthly expenditures was the energy bill? (   )

(4) What percentage of the monthly expenditures was the clothing cost? (   )

(5) How many times more did housing cost per month than clothing? (   )

4 The band graph below shows the ratio of different activities in Stuart's daily life.

4 points per question

Stuart's daily activities

Studying	Free time	Eating	Chores

0 10 20 30 40 50 60 70 80 90 100%

(1) What percentage of his day does Stuart spend studying? (   )

(2) What percentage of his day does Stuart spend as free time? (   )

(3) What percentage of his day does Stuart spend doing chores? (   )

(4) What percentage of his day does Stuart spend eating? (   )

(5) What is the ratio of the amount of time Stuart spends on chores compared to the amount of free time (   )
he has?

Graphs are fun, right? Good job.

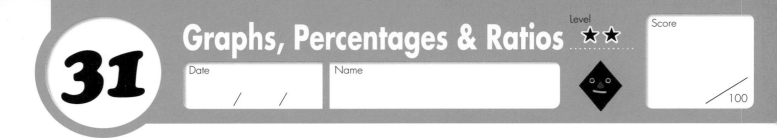

31

Graphs, Percentages & Ratios

Level ★★

Date / /

Name

Score
/100

1 The circle graph to the right shows the percentages of the different types of farm animals in a small town.

5 points per question

(1) What percentage of the whole are cattle for beef? (60%)

(2) What percentage of the whole are cows? ( )

(3) What percentage of the whole are pigs? ( )

(4) What percentage of the whole are other? ( )

(5) How many times more cow is there than pig? ( )

(6) How many times more pig is there than cow? ( )

Types of farm animals

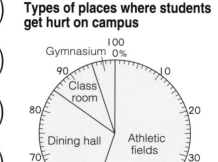

2 The circle graph to the right shows the percentages of the different places where students get hurt on campus.

5 points per question

(1) What percentage of students were injured on the athletic fields? ( )

(2) What percentage of students were injured in the dining hall? ( )

(3) What percentage of students were injured in a classroom? ( )

(4) What percentage of students were injured in the gymnasium? ( )

(5) How many times more students were hurt in the dining hall than in the gymnasium? ( )

(6) How many times more students were hurt in the classrooms than in the dining hall? ( )

Types of places where students get hurt on campus

3 The circle graph on the right shows the percentages of the different types of occupations in a certain city.

4 points per question

(1) What percentage of the occupations are industrial? (   )

(2) What percentage of the occupations are commercial? (   )

(3) What percentage of the occupations are agricultural? (   )

(4) What percentage of the occupations are other? (   )

(5) How many times more occupations are commercial compared to agricultural? (   )

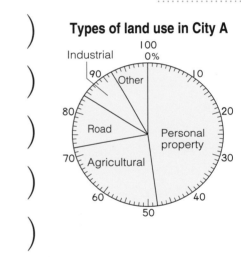

Types of occupations

4 The circle graph on the right shows the percentages of the different types of land use in a certain city.

4 points per question

(1) What percentage of the land is used for personal property? (   )

(2) What percentage of the land is used for agricultural work? (   )

(3) What percentage of the land is used for road? (   )

(4) What percentage of the land is used for industrial work? (   )

(5) How many times more land is used for industrial work compared to personal property? (   )

Types of land use in City A

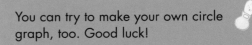

You can try to make your own circle graph, too. Good luck!

63

32 Graphs, Percentages & Ratios

Level ★★

Date / /

Name

Score

/100

1 Mrs. McPhee had all the children sit randomly in different zones in class. The table on the right shows how the children ended up sitting.

10 points per question

(1) Use the number sentences below to calculate the percentage of children sitting in each zone.

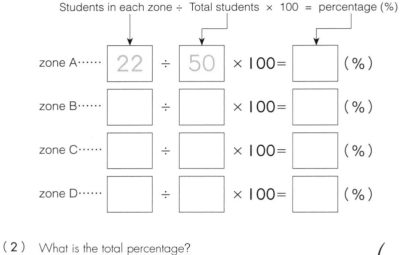

Students in each zone ÷ Total students × 100 = percentage (%)

zone A⋯⋯ 22 ÷ 50 × 100 = ☐ (%)

zone B⋯⋯ ☐ ÷ ☐ × 100 = ☐ (%)

zone C⋯⋯ ☐ ÷ ☐ × 100 = ☐ (%)

zone D⋯⋯ ☐ ÷ ☐ × 100 = ☐ (%)

Students per sitting zone

Zone	Number of people	Percentage
A	22	( )
B	13	( )
C	8	( )
D	7	( )
Total	50	( )

(2) What is the total percentage?

( )

(3) Fill the table with the percentages you calculated.

2 Calculate the percentages in order to fill the tables below. Round each percentage to the nearest ones place.

10 points per question

(1) **Types of rides chosen**

Ride	Number of people	Percentage
Roller coaster	36	(45)
Bumper Cars	24	( )
Merry-Go-Round	12	( )
Parachute	8	( )
Total	80	( )

(2) **Living zone populations**

Zone	Number of people	Percentage
A	22	( )
B	21	( )
C	13	( )
D	9	( )
Total	65	( )

3 The school nurse made the table on the right in order to tabulate the types of places her students were injured.

10 points per question

(1) Use the number sentences below, calculate the percentages in order to fill the table.

Fields .... ☐ ÷ ☐ × 100 = ☐ (%)

Gym .... ☐ ÷ ☐ × 100 = ☐ (%)

Cafeteria .... ☐ ÷ ☐ × 100 = ☐ (%)

Class .... ☐ ÷ ☐ × 100 = ☐ (%)

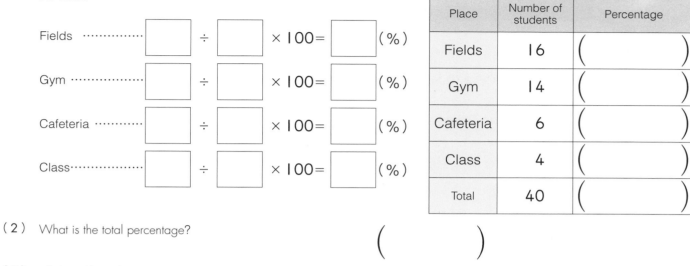

Places students were injured

Place	Number of students	Percentage
Fields	16	( )
Gym	14	( )
Cafeteria	6	( )
Class	4	( )
Total	40	( )

(2) What is the total percentage?

( )

(3) Fill the table with the percentages you calculated.

4 Calculate the percentages in order to fill the tables below. Round each percentage to the nearest ones place.

10 points per question

(1) **Types of drinks ordered**

Drink	Number of people	Percentage
Coffee	30	( )
Tea	21	( )
Juice	17	( )
Cocoa	12	( )
Total	80	( )

(2) **Types of books bought**

Book	Number of people	Percentage
Comic book	141	( )
Novel	104	( )
Non-Fiction	62	( )
Art book	28	( )
Magazine	15	( )
Total	350	( )

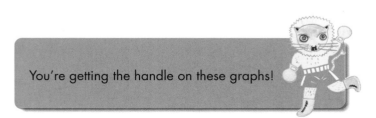

You're getting the handle on these graphs!

Date / /

Name

Score

/100

1 Use the tables pictured here to fill in the band graphs on the right.

10 points per question

(1)

Types of land use in Town A

Residential	Agricultural	Forest	Other
40%	30%	20%	10%

Types of land use in Town A

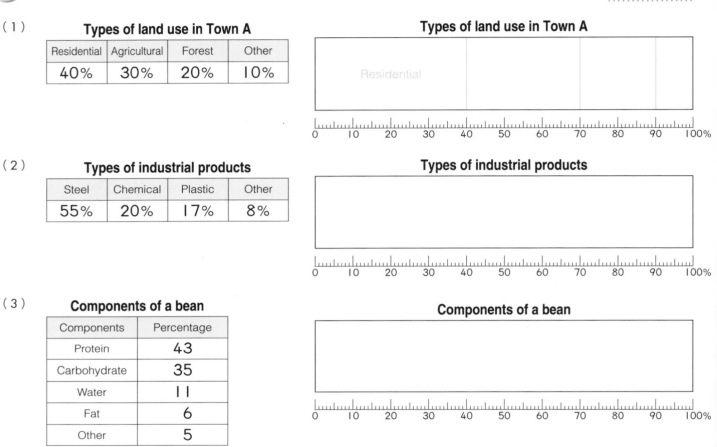

Residential

0 10 20 30 40 50 60 70 80 90 100%

(2)

Types of industrial products

Steel	Chemical	Plastic	Other
55%	20%	17%	8%

Types of industrial products

0 10 20 30 40 50 60 70 80 90 100%

(3)

Components of a bean

Components	Percentage
Protein	43
Carbohydrate	35
Water	11
Fat	6
Other	5

Components of a bean

0 10 20 30 40 50 60 70 80 90 100%

2 The table on the right shows the type of books in our classroom.

10 points per question

(1) Calculate the percentages in order to fill the table.

Types of books in our classroom

Type	Number of books	Percentage
Novel	47	( )
Science	42	( )
Biography	27	( )
Other	14	( )
Total	130	( )

(2) Fill the band graph below by using the table on the right.

Types of books in our classroom

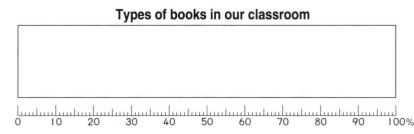

0 10 20 30 40 50 60 70 80 90 100%

3 Use the tables pictured here to fill in the circle graphs below.

10 points per question

（1）
Types of vehicles in the street

Car	Truck	Taxi	Other
40%	30%	20%	10%

（2）
Types of expenditures

Snacks	Stationery	Books	Other
65%	20%	12%	3%

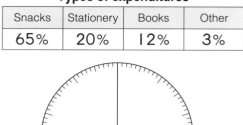

（3）
Proportion of export value by type of product

Product	Percentage
Pharmaceuticals	41
Vehicles	17
Steel	11
Precision instruments	5
Other	26

Proportion of export value by type of product

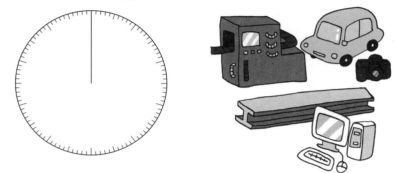

4 The table on the right shows how many of each type of store are in Town B.

10 points per question

（1）Calculate the percentages in order to fill the table on the right.

（2）Use the table on the right to fill the circle graph below.

Types of stores in Town B

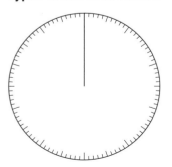

Types of stores in Town B

Store type	Number of stores	Percentage
Clothing	29	( )
Grocery	23	( )
Hardware	16	( )
Furniture	3	( )
Other	19	( )
Total	90	( )

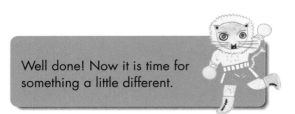

Well done! Now it is time for something a little different.

Mixed Problems

34

Date   /   /

Name

Level ★★★

Score

/100

1 You used coins to arrange the following squares with sides of 2 coins, 3 coins and 4 coins.

10 points per question

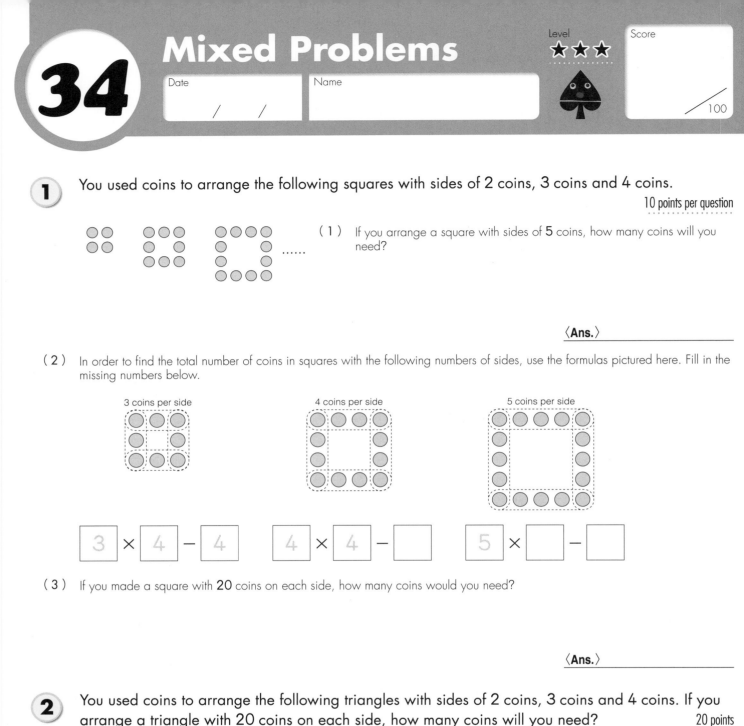

(1) If you arrange a square with sides of **5** coins, how many coins will you need?

⟨**Ans.**⟩ _____

(2) In order to find the total number of coins in squares with the following numbers of sides, use the formulas pictured here. Fill in the missing numbers below.

3 coins per side

4 coins per side

5 coins per side

$3 \times 4 - 4$ $4 \times 4 - \boxed{}$ $5 \times \boxed{} - \boxed{}$

(3) If you made a square with **20** coins on each side, how many coins would you need?

⟨**Ans.**⟩ _____

2 You used coins to arrange the following triangles with sides of 2 coins, 3 coins and 4 coins. If you arrange a triangle with 20 coins on each side, how many coins will you need?

20 points

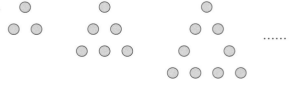

⟨**Ans.**⟩ _____

3 On the cards below, the number of circles is increasing at a regular rate. How many circles will there be on 20th card?

10 points

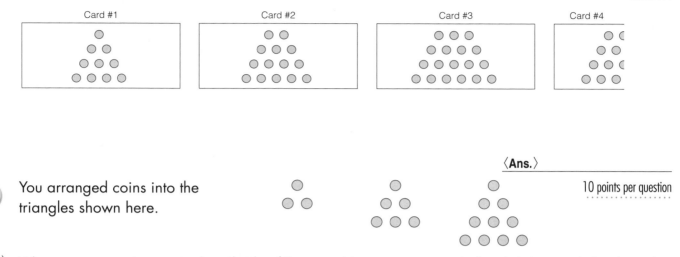

Card #1 Card #2 Card #3 Card #4

⟨**Ans.**⟩ _____

4 You arranged coins into the triangles shown here.

10 points per question

(1) When you arrange coins into triangles with sides of **3** coins and **4** coins, you can use the formula below to calculate the number of coins needed. Fill in the missing numbers below.

3 coins per side

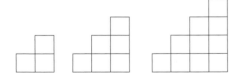

3 rows

3 1

4 coins per side

$$\left(3 + \boxed{1}\right) \times \boxed{3} \div \boxed{2}$$

$$\left(4 + \boxed{1}\right) \times \boxed{} \div \boxed{}$$

(2) If you arrange the coins into a triangle with **20** per side, how many coins will you need?

⟨**Ans.**⟩ _____

5 You arranged some square boards into the shapes shown below. How many square boards do you need in order to make a similar shape with 20 squares on the bottom line?

20 points

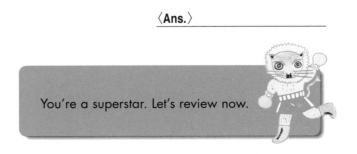

⟨**Ans.**⟩ _____

You're a superstar. Let's review now.

Review

35

Date   /   /

Name

Level ★★★

Score
/100

---

1 If 1 foot of ribbon costs 180¢, how much does 0.8 foot of ribbon cost?   10 points

⟨Ans.⟩ _____

2 The mechanic has 32.4 liters of oil in the back. If he divides the oil into 1.8-liter bottles evenly, how many bottles will he need?   10 points

⟨Ans.⟩ _____

3 The corner store has a double pack with $\frac{5}{7}$ liter of regular milk and $\frac{2}{7}$ liter of non-fat milk. How much milk is in the pack altogether?   10 points

⟨Ans.⟩ _____

4 Chen is choosing between two boards for his birdhouse. One is 8 feet long and the other is 9 feet long. How many times longer is the 9-foot long board? Answer in decimals.   10 points

⟨Ans.⟩ _____

5 The school supply store got 500 notebooks on Friday. By next Monday, the store sold 20% of the notebooks. How many notebooks did they sell over the weekend?   10 points

⟨Ans.⟩ _____

6 The corner store gets ice cream bars for $3.20 each. If they want to sell them for 25% more, what price should they put on the bars?

10 points

〈**Ans.**〉 _____

7 I started painting the garage today. Before I started, I had $\frac{5}{7}$ liter of paint. If I had $\frac{3}{7}$ liter of paint when I was done, how much paint did I use?

10 points

〈**Ans.**〉 _____

8 Ken runs 2.5 kilometers every day. If he runs every day for 30 days, how far will he have run in all?

10 points

〈**Ans.**〉 _____

9 If you divide 5.6 meters of tape into 3 pieces, how long is each piece? Round the answer to the nearest tenth.

10 points

〈**Ans.**〉 _____

10 At the construction site, there's a 1-yard piece of pipe that weighs 2.3 ounces. How much does 4.5 yards of that pipe weigh?

10 points

〈**Ans.**〉 _____

Almost there!

1 1.2 meters of my hose weighs 1.5 kilograms. How much does 1 meter of my hose weigh?　　10 points

〈Ans.〉 _____

2 From Sally's house to the train station is 2.8 kilometers. That is 3.5 times as long as the distance from her house to the bus stop. How far is it from her house to the bus stop?　　10 points

〈Ans.〉 _____

3 Of the 20 people on Tina's soccer team, 6 people are fifth graders. What ratio is the number of fifth graders to the whole?　　10 points

〈Ans.〉 _____

4 Mike went hiking and walked 9 kilometers. That distance was 60% of the whole. How far does Mike have to hike today?　　10 points

〈Ans.〉 _____

5 You arranged a group of black and white stones as shown below. Is the 18th stone from the left white or black?　　10 points

○ ● ○ ● ○ ● ○ ● ......

〈Ans.〉 _____

6 If you go east from Ted's house for $\frac{4}{8}$ kilometer, there is a bus station. $\frac{7}{8}$ kilometer west of his house there is a train station. Which station is further away, and how much further away is that station?

10 points

〈**Ans.**〉 _____

7 We took 50 meters of rope and divided it into 1.6-meter sections. How many sections did we make and how much was left?

10 points

〈**Ans.**〉 _____

8 The soccer team in Littleton drew 28,680 people last week to their game, and 25,760 people this week. What was the difference between the number of people that came to each game? Round the answer to the nearest thousands place.

10 points

〈**Ans.**〉 _____

9 Jane is looking at a cap that cost $28 last week. If it is on sale at 25% off, how much will it cost now?

10 points

〈**Ans.**〉 _____

10 Our art class has some red sheets of paper and some blue sheets of paper for class today. In all, we have 36 sheets of paper. If the number of blue sheets is 3 times as much as the number of red sheets, how many red and blue sheets do we have?

10 points

〈**Ans.**〉 Red _____ Blue _____

You did it. Congratulations!

1 Review pp 2,3

1. $252 \div 6 = 42$ **Ans.** 42 boxes
2. $\$5 = 500¢, \ 500 \div 4 = 125$ **Ans.** 125¢
3. $68 \div 14 = 4 \ R \ 12$ **Ans.** 4 sets, 12 books remain
4. $1.2 + 2.5 = 3.7$ **Ans.** 3.7 kg
5. $1.2 - 0.3 = 0.9$ **Ans.** 0.9 lb
6. $120 \div 3 = 40$ **Ans.** 40 L
7. $24 \div 8 = 3$ **Ans.** 3 times
8. $260 \div 35 = 7 \ R \ 15$ **Ans.** 7 people, 15 flyers remain
9. $500 - 70 \times 6 = 80$ **Ans.** 80¢
10. $38 - 22 = 16, \ 16 \div 2 = 8$ **Ans.** $8

2 Review pp 4,5

1. $150 \div 5 = 30$ **Ans.** 30 flyers
2. $1.5 + 1.2 = 2.7$ **Ans.** 2.7 L
3. $90 \div (5 \times 6) = 3$ **Ans.** 3 t-shirts
4. $4 \times 5 + 6 \times 3 = 38$ **Ans.** $38
5. $90 \div 7 = 12 \ R \ 6$ **Ans.** 12 pieces, 6 cm remain
6. $68 \div 5 = 13 \ R \ 3 \quad 13 + 1 = 14$ **Ans.** 14 trips
7. $2.5 - 1.2 = 1.3$ **Ans.** 1.3 L
8. $365 \div 7 = 52 \ R \ 1$ **Ans.** 52 weeks, 1 day remains
9. $500 \div 80 = 6 \ R \ 20$ **Ans.** 6 ice creams, 20¢ remain
10. $200 \div (25 + 15) = 200 \div 40 = 5$ **Ans.** 5 sets

3 Round Numbers pp 6,7

1. $20,000 + 30,000 = 50,000$ **Ans.** 50,000 people
2. $30,000 + 30,000 = 60,000$ **Ans.** 60,000 people
3. $130,000 + 110,000 = 240,000$ **Ans.** 240,000 people
4. $40,000 + 60,000 = 100,000$ **Ans.** $100,000
5. $1,000 + 3,000 = 4,000$ **Ans.** $4,000
6. $36,000 + 43,000 = 79,000$ **Ans.** 79,000 km
7. $416,000 + 383,000 = 799,000$ **Ans.** $799,000¢
8. $124,000 + 268,000 = 392,000$ **Ans.** 392,000 ft.²
9. $18,000 + 22,000 = 40,000$ **Ans.** 40,000 bottles
10. $27,000 + 3,000 = 30,000$ **Ans.** 30,000 yd.

4 Round Numbers pp 8,9

1. $32,000 - 28,000 = 4,000$ **Ans.** 4,000 fans
2. $30,000 - 20,000 = 10,000$ **Ans.** 10,000 people
3. $130,000 - 110,000 = 20,000$ **Ans.** 20,000 men
4. $60,000 - 40,000 = 20,000$ **Ans.** $20,000
5. $3,000 - 2,000 = 1,000$ **Ans.** 1,000 people
6. $42,000 - 36,000 = 6,000$ **Ans.** 6,000 mi.
7. $403,000 - 398,000 = 5,000$ **Ans.** $5,000
8. $268,000 - 174,000 = 94,000$ **Ans.** 94,000 m²
9. $22,000 - 18,000 = 4,000$ **Ans.** 4,000 widgets
10. $25,000 - 3,000 = 22,000$ **Ans.** 22,000 m

5 Odd & Even pp 10,11

1 (1)

Number of times across the threshold	0	1	2	3	4
Where Vicki is	room	hall	room	hall	room

(2) Even
(3) Odd
(4) Classroom

2 (1) Odd
(2) Even
(3) Girl

3 Black

4 Yellow

5 Blue

6 (1) Boy
(2) Girl

6 Fractions pp 12,13

1. $\frac{1}{4} + \frac{2}{4} = \frac{3}{4}$ **Ans.** $\frac{3}{4}$ L
2. $\frac{3}{6} + \frac{2}{6} = \frac{5}{6}$ **Ans.** $\frac{5}{6}$ ft.²
3. $\frac{2}{5} + \frac{1}{5} = \frac{3}{5}$ **Ans.** $\frac{3}{5}$
4. $\frac{2}{7} + \frac{3}{7} = \frac{5}{7}$ **Ans.** $\frac{5}{7}$ m
5. $\frac{1}{4} + \frac{3}{4} = 1$ **Ans.** 1 lb.
6. $\frac{3}{7} + \frac{5}{7} = \frac{8}{7}$ **Ans.** $\frac{8}{7}$ L $\left(1\frac{1}{7} \text{ L}\right)$
7. $\frac{5}{8} + \frac{6}{8} = \frac{11}{8}$ **Ans.** $\frac{11}{8}$ lb. $\left(1\frac{3}{8} \text{ lb.}\right)$
8. $\frac{4}{9} + \frac{7}{9} = \frac{11}{9}$ **Ans.** $\frac{11}{9}$ km $\left(1\frac{2}{9} \text{ km}\right)$
9. $\frac{2}{8} + \frac{1}{8} = \frac{3}{8}$ **Ans.** $\frac{3}{8}$ L
10. $\frac{3}{4} + \frac{2}{4} = \frac{5}{4}$ **Ans.** $\frac{5}{4}$ yd. $\left(1\frac{1}{4} \text{ yd.}\right)$

7 Fractions

pp 14,15

1) $\frac{3}{5} - \frac{2}{5} = \frac{1}{5}$ **Ans.** $\frac{1}{5}$ L

2) $\frac{3}{4} - \frac{1}{4} = \frac{2}{4}$ **Ans.** $\frac{2}{4}$ L

3) $\frac{4}{5} - \frac{1}{5} = \frac{3}{5}$ **Ans.** $\frac{3}{5}$ mi.

4) $\frac{3}{7} - \frac{2}{7} = \frac{1}{7}$

       Ans. $\frac{1}{7}$ L more today.

5) $\frac{8}{9} - \frac{5}{9} = \frac{3}{9}$

       Ans. City hall is $\frac{3}{9}$ mile further away.

6) $\frac{6}{7} - \frac{2}{7} = \frac{4}{7}$ **Ans.** $\frac{4}{7}$ L

7) $1 - \frac{3}{4} = \frac{1}{4}$ **Ans.** $\frac{1}{4}$ lb.

8) $\frac{7}{8} - \frac{3}{8} = \frac{4}{8}$ **Ans.** $\frac{4}{8}$ L

9) $1 - \frac{3}{5} = \frac{2}{5}$ **Ans.** $\frac{2}{5}$ km

10) $\frac{5}{7} - \frac{2}{7} = \frac{3}{7}$ **Ans.** $\frac{3}{7}$

8 Fractions

pp 16,17

1) $2 \div 3 = \frac{2}{3}$ **Ans.** $\frac{2}{3}$ m

2) $3 \div 4 = \frac{3}{4}$ **Ans.** $\frac{3}{4}$ m

3) $5 \div 3 = \frac{5}{3}$ **Ans.** $\frac{5}{3}$ lb. $\left(1\frac{2}{3}\text{ lb.}\right)$

4) $3 \div 7 = \frac{3}{7}$, $4 \div 7 = \frac{4}{7}$, $\frac{4}{7} - \frac{3}{7} = \frac{1}{7}$

       Ans. Girls got $\frac{1}{7}$ liter more.

5) $3 \div 3 = 1$, $4 \div 5 = \frac{4}{5}$, $1 - \frac{4}{5} = \frac{1}{5}$

       Ans. Boys got $\frac{1}{5}$ foot more.

6) $3 \div 5 = \frac{3}{5}$ **Ans.** $\frac{3}{5}$ times

7) $6 \div 7 = \frac{6}{7}$ **Ans.** $\frac{6}{7}$ times

8) $8 \div 9 = \frac{8}{9}$ **Ans.** $\frac{8}{9}$ times

9) $3 \div 8 = \frac{3}{8}$ **Ans.** $\frac{3}{8}$ times

10) $8 \div 15 = \frac{8}{15}$ **Ans.** $\frac{8}{15}$ times

9 Decimals

pp 18,19

1) $0.2 \times 7 = 1.4$ **Ans.** 1.4 L

2) $0.4 \times 3 = 1.2$ **Ans.** 1.2 m

3) $0.8 \times 6 = 4.8$ **Ans.** 4.8 km

4) $0.5 \times 5 = 2.5$ **Ans.** 2.5 lb.

5) $0.4 \times 6 = 2.4$ **Ans.** 2.4 ft.

6) $3.2 \times 4 = 12.8$ **Ans.** 12.8 m

7) $4.5 \times 2 = 9$ **Ans.** 9 mi.

8) $2.8 \times 5 = 14$ **Ans.** 14 lb.

9) $1.6 \times 25 = 40$ **Ans.** 40 km

10) $2.5 \times 18 = 45$ **Ans.** 45 ft.

10 Decimals

pp 20,21

1) $1.2 \div 3 = 0.4$ **Ans.** 0.4 L

2) $8.5 \div 5 = 1.7$ **Ans.** 1.7 m

3) $6.4 \div 4 = 1.6$ **Ans.** 1.6 lb.

4) $7.8 \div 6 = 1.3$ **Ans.** 1.3 m

5) $9.2 \div 4 = 2.3$ **Ans.** 2.3 ft.

6) $5.4 \div 3 = 1.8$ **Ans.** 1.8 L

7) $5.4 \div 6 = 0.9$ **Ans.** 0.9 ft.

8) $10.4 \div 5 = 2.6$ **Ans.** 2.6 kg

9) $7.5 \div 5 = 1.5$ **Ans.** 1.5 km

10) $15.4 \div 7 = 2.2$ **Ans.** 2.2 lb.

11 Decimals

pp 22,23

1) $8.5 \div 3 = 2 \text{ R } 2.5$

       Ans. 2 flowers, 2.5 ft. remain

2) $13.5 \div 2 = 6 \text{ R } 1.5$

       Ans. 6 pipes, 1.5 ft. remain

3) $7.5 \div 2 = 3 \text{ R } 1.5$

       Ans. 3 cakes, 1.5 lb. remain

4) $7.8 \div 3 = 2 \text{ R } 1.8$

       Ans. 2 necklaces, 1.8 ft. remain

5) $9.5 \div 4 = 2 \text{ R } 1.5$

       Ans. 2 pieces, 1.5 ft. remain

6) $5.4 \div 3 = 1 \text{ R } 2.4$

       Ans. 1 bottle, 2.4 L remain

7) $9.4 \div 4 = 2 \text{ R } 1.4$

       Ans. 2 pieces, 1.4 m remain

8) $18.2 \div 5 = 3 \text{ R } 3.2$

       Ans. 3 bags, 3.2 kg remain

9) $24.4 \div 7 = 3 \text{ R } 3.4$

       Ans. 3 friends, 3.4 m remain

10) $19.5 \div 3 = 6 \text{ R } 1.5$

       Ans. 6 baskets, 1.5 kg remain

12 Decimals

1. $3.4 \div 5 = 0.68$ **Ans.** 0.68 ft.
2. $7.4 \div 5 = 1.48$ **Ans.** 1.48 ft.
3. $2.6 \div 4 = 0.65$ **Ans.** 0.65 L
4. $1.5 \div 6 = 0.25$ **Ans.** 0.25 L
5. $6.6 \div 4 = 1.65$ **Ans.** 1.65 ft.
6. $9 \div 6 = 1.5$ **Ans.** 1.5 kg
7. $15 \div 4 = 3.75$ **Ans.** 3.75 m²
8. $7 \div 5 = 1.4$ **Ans.** 1.4 km
9. $6 \div 4 = 1.5$ **Ans.** 1.5 kg
10. $21 \div 6 = 3.5$ **Ans.** 3.5 kg

13 Decimals

1. $8.3 \div 6 = 1.38$ **Ans.** 1.4 lb.
2. $2.5 \div 6 = 0.41$ **Ans.** 0.4 L
3. $16.3 \div 3 = 5.43$ **Ans.** 5.4 ft.
4. $7.6 \div 3 = 2.53$ **Ans.** 2.5 ft.
5. $8.2 \div 9 = 0.91$ **Ans.** 0.9 lb.
6. $5.4 \div 8 = 0.67$ **Ans.** 0.7 m²
7. $6.5 \div 3 = 2.16$ **Ans.** 2.2 km
8. $9.5 \div 4 = 2.37$ **Ans.** 2.4 m
9. $4.4 \div 3 = 1.46$ **Ans.** 1.5 kg
10. $1.7 \div 8 = 0.21$ **Ans.** 0.2 kg

14 Decimals

1. $45 \div 30 = 1.5$ **Ans.** 1.5 times
2. $15 \div 30 = 0.5$ **Ans.** 0.5 times
3. $56 \div 35 = 1.6$ **Ans.** 1.6 times
4. $35 \div 56 = 0.625$ **Ans.** 0.625 times
5. $15 \div 4 = 3.75$ **Ans.** 3.75 times
6. $2.5 \div 5 = 0.5$ **Ans.** 0.5 times
7. $5.4 \div 9 = 0.6$ **Ans.** 0.6 times
8. $7 \div 5 = 1.4$ **Ans.** 1.4 times
9. $5.4 \div 3 = 1.8$ **Ans.** 1.8 times
10. $1.7 \div 2 = 0.85$ **Ans.** 0.85 times

15 Decimals

1. $250 \times 3.6 = 900$ **Ans.** 900 oz.
2. $3 \times 2.5 = 7.5$ **Ans.** $7.5
3. $7 \times 1.5 = 10.5$ **Ans.** $10.5
4. $15 \times 1.8 = 27$ **Ans.** 27 lb.

5. $8 \times 0.6 = 4.8$ **Ans.** $4.8
6. $1.2 \times 2.5 = 3$ **Ans.** 3 kg
7. $9.5 \times 6.5 = 61.75$ **Ans.** 61.75 km
8. $25.4 \times 1.5 = 38.1$ **Ans.** 38.1 m²
9. $2.6 \times 3.4 = 8.84$ **Ans.** 8.84 kg
10. $1.2 \times 0.8 = 0.96$ **Ans.** 0.96 kg

16 Decimals

1. $38 \times 1.5 = 57$ **Ans.** 57 lb.
2. $450 \times 1.2 = 540$ **Ans.** 540 m
3. $16 \times 2.5 = 40$ **Ans.** 40 lb.
4. $12 \times 1.4 = 16.8$ **Ans.** 16.8 L
5. $3 \times 0.5 = 1.5$ **Ans.** 1.5 hours
6. $9 \div 1.8 = 5$ **Ans.** $5
7. $10 \div 2.5 = 4$ **Ans.** $4
8. $6 \div 2.5 = 2.4$ **Ans.** $2.4
9. $36 \div 1.8 = 20$ **Ans.** 20 bottles
10. $306 \div 0.9 = 340$ **Ans.** 340 g

17 Decimals

1. $6.4 \div 1.6 = 4$ **Ans.** 4 lb.
2. $4.2 \div 3.5 = 1.2$ **Ans.** 1.2 lb.
3. $2.7 \div 4.5 = 0.6$ **Ans.** 0.6 lb.
4. $4.2 \div 3.5 = 1.2$ **Ans.** 1.2 L
5. $1.4 \div 2.8 = 0.5$ **Ans.** 0.5 lb.
6. $28.8 \div 1.2 = 24$ **Ans.** 24 watermelons
7. $37.5 \div 1.5 = 25$ **Ans.** 25 bags
8. $77.5 \div 2.5 = 31$ **Ans.** 31 days
9. $13.4 \div 2.5 = 5 \, R \, 0.9$ **Ans.** 5 pieces, 0.9 m remains
10. $21.9 \div 1.8 = 12 \, R \, 0.3$ **Ans.** 12 bottles, 0.3 L milk remains

18 Decimals

1. $8.7 \div 5.8 = 1.5$ **Ans.** 1.5 times
2. $104 \div 6.5 = 16$ **Ans.** 16 times
3. $28.5 \div 1.5 = 19$ **Ans.** 19 times
4. $12.9 \div 8.6 = 1.5$ **Ans.** 1.5 times
5. $4.3 \div 8.6 = 0.5$ **Ans.** 0.5 times
6. $54.6 \div 1.5 = 36.4$ **Ans.** 36.4 kg
7. $4.5 \div 1.8 = 2.5$ **Ans.** 2.5 m
8. $168.6 \div 1.2 = 140.5$ **Ans.** 140.5 cm
9. $28.5 \div 1.5 = 19$ **Ans.** 19 kg
10. $5.4 \div 1.2 = 4.5$ **Ans.** 4.5 L

19 Decimals
pp 38, 39

1 (1) $8 \div 10 = 0.8$ **Ans.** 0.8 lb.
(2) $0.8 \times 1.8 = 1.44$ **Ans.** 1.44 lb.

2 $21 \div 30 \times 70 = 49$ **Ans.** 49 oz.

3 $45.3 \div 1.5 \times 5.5 = 166.1$ **Ans.** 166.1 oz.

4 $12.4 \div 0.8 \times 22.6 = 350.3$ **Ans.** 350.3 ft.

5 $1.2 \div 0.6 \times 2.5 = 5$ **Ans.** 5 L

6 $(18.2 - 1.4) \div 1.2 = 14$ **Ans.** 14 dictionaries

7 $(20.9 - 1.1) \div 1.8 = 11$ **Ans.** 11 bottles

8 $(10 - 0.2) \div 12.5 = 0.784$ **Ans.** 0.784 kg

9 $(10.6 + 15.8) \times 5.5 = 145.2$ **Ans.** 145.2 L

10 $8.6 \times (12.5 + 20.8) = 286.38$ **Ans.** 286.38 km

20 Ratios
pp 40, 41

1 (1) Boys $7 \div 10 = 0.7$, Girls $3 \div 10 = 0.3$
(2) Boys $3 \div 5 = 0.6$, Girls $2 \div 5 = 0.4$
(3) Boys $1 \div 5 = 0.2$, Girls $4 \div 5 = 0.8$
(4) Boys $1 \div 4 = 0.25$, Girls $3 \div 4 = 0.75$
(5) Boys $6 \div 15 = 0.4$, Girls $9 \div 15 = 0.6$

2 (1) 4 %   (2) 9 %   (3) 16 %   (4) 45 %
(5) 50 %   (6) 100 %   (7) 109 %   (8) 200 %
(9) 0.6 %   (10) 80.7 %

3 (1) 0.03   (2) 0.09   (3) 0.12   (4) 0.6
(5) 0.97   (6) 1   (7) 2.5   (8) 3.07
(9) 0.008   (10) 0.602

21 Ratios
pp 42, 43

1 $16 \div 20 = 0.8$ **Ans.** 0.8 times

2 $16 \div 20 = 0.8$ **Ans.** 0.8

3 $7 \div 35 = 0.2$ **Ans.** 0.2

4 $35 \div 40 = 0.875$ **Ans.** 0.875

5 $18 \div 36 = 0.5$ **Ans.** 0.5

6 $9 \div 150 = 0.06$ **Ans.** 0.06

7 $15 \div 20 = 0.75$ **Ans.** 0.75

8 $40 \div 200 = 0.2$ **Ans.** 0.2

9 $20 \div 500 = 0.04$ **Ans.** 0.04

10 $64.8 \div 13.5 = 4.8$ **Ans.** 4.8

Advice
You can also express your ratios as fractions.

22 Ratios
pp 44, 45

1 $35 \times 1.2 = 42$ **Ans.** 42 lb.

2 $1.4 \times 1.5 = 2.1$ **Ans.** 2.1 ft.

3 $35 \times 0.9 = 31.5$ **Ans.** 31.5 lb.

4 $180 \times 0.6 = 108$ **Ans.** 108 boys

5 $48 \times 0.7 = 33.6$ **Ans.** 33.6 L

6 $240 \times 0.3 = 72$ **Ans.** 72 pages

7 $860 \times 0.2 = 172$ **Ans.** 172 students

8 $140 \times 0.8 = 112$ **Ans.** 112 cm

9 $6,855 \times 1.2 = 8,226$ **Ans.** 8,226 people

10 $400 \times 0.1 = 40$ **Ans.** 40 apples

23 Ratios
pp 46, 47

1 $49 \div 1.4 = 35$ **Ans.** 35 lb.

2 $49 \div 0.7 = 70$ **Ans.** 70 lb.

3 $28 \div 0.4 = 70$ **Ans.** 70 games

4 $42 \div 0.6 = 70$ **Ans.** 70 problems

5 $4.8 \div 0.2 = 24$ **Ans.** 24 L

6 $24 \div 0.2 = 120$ **Ans.** 120 students

7 $204 \div 0.8 = 255$ **Ans.** 255 pages

8 $306 \div 0.4 = 765$ **Ans.** 765 children

9 $136 \div 0.8 = 170$ **Ans.** 170 cm

10 $2,520 \div 1.4 = 1,800$ **Ans.** $1,800

24 Ratios
pp 48, 49

1 $16 \div 40 = 0.4$ **Ans.** 0.4

2 $24 \div 40 = 0.6$ **Ans.** 60 %

3 $18 \div 60 = 0.3$ **Ans.** 30 %

4 $45 \div 50 = 0.9$ **Ans.** 90 %

5 $30 \div 150 = 0.2$ **Ans.** 20 %

6 $900 \div 3,600 = 0.25$ **Ans.** 25 %

7 $3.6 \div 24 = 0.15$ **Ans.** 15 %

8 $0.5 \div 4 = 0.125$ **Ans.** 12.5 %

9 $321 \div 6,420 = 0.05$ **Ans.** 5 %

10 $4 \div 32 = 0.125$ **Ans.** 12.5 %

25 Ratios
pp 50, 51

1 $35 \times 0.2 = 7$ **Ans.** 7 students

2 $600 \times 0.3 = 180$ **Ans.** 180 magazines

3 $80 \times 0.6 = 48$ **Ans.** 48 novels

4 $45 \times 0.2 = 9$ **Ans.** $9

5 $24 \times 0.4 = 9.6$ **Ans.** 9.6 ft.^{2}

6 $120 \times 0.7 = 84$ **Ans.** 84 people

7 $200 \times 0.2 = 40$ **Ans.** $40

8 $150 \times 1.2 = 180$ **Ans.** 180 lb.

9 $640 \times 0.05 = 32$ **Ans.** 32 students

10 $180 \times 0.86 = 154.8$ **Ans.** 154.8 g

26 Ratios
pp 52,53

1. $12 \div 0.1 = 120$ **Ans.** 120 people
2. $21 \div 0.6 = 35$ **Ans.** 35 people
3. $6 \div 0.2 = 30$ **Ans.** 30 L
4. $12 \div 0.3 = 40$ **Ans.** 40 books
5. $3 \div 0.4 = 7.5$ **Ans.** $7.5
6. $80 \div 0.1 = 800$ **Ans.** 800 m²
7. $240 \div 0.4 = 600$ **Ans.** 600 children
8. $3,600 \div 1.2 = 3,000$ **Ans.** 3,000 kg
9. $41.8 \div 1.1 = 38$ **Ans.** 38 kg
10. $140 \div 0.1 = 1,400$ **Ans.** 1,400 tomatoes

27 Ratios
pp 54,55

1. $10 \div (90 + 10) = 0.1$ **Ans.** 10 %
2. $30 \div (120 + 30) = 0.2$ **Ans.** 20 %
3. $20 \div (480 + 20) = 0.04$ **Ans.** 4 %
4. $45 \div (45 + 5) = 0.9$ **Ans.** 90 %
5. $62 \div (62 + 18) = 0.775$ **Ans.** 77.5 %
6. $(500 + 50) \div 500 = 1.1$ **Ans.** 110 %
7. $(300 + 60) \div 300 = 1.2$ **Ans.** 120 %
8. $(456 + 114) \div 456 = 1.25$ **Ans.** 125 %
9. $(250 + 150) \div 250 = 1.6$ **Ans.** 160 %
10. $(2,800 + 600) \div 2,800 = 1.214\cdots$ **Ans.** 121 %

28 Ratios
pp 56,57

1. $(300 - 270) \div 300 = 0.1$ **Ans.** 10 %
2. $(200 - 180) \div 200 = 0.1$ **Ans.** 10 %
3. $(500 - 400) \div 500 = 0.2$ **Ans.** 20 %
4. $(800 - 380) \div 800 = 0.525$ **Ans.** 52.5 %
5. $(250 - 200) \div 250 = 0.2$ **Ans.** 20 %
6. $60 \div (560 - 60) = 0.12$ **Ans.** 12 %
7. $40 \div (540 - 40) = 0.08$ **Ans.** 8 %
8. $8 \div (48 - 8) = 0.2$ **Ans.** 20 %
9. $15 \div (75 - 15) = 0.25$ **Ans.** 25 %
10. $300 \div (4,300 - 300) = 0.075$ **Ans.** 7.5 %

29 Ratios
pp 58,59

1. $300 \times 1.2 = 360$ **Ans.** $360
2. $400 \times 1.1 = 440$ **Ans.** $440
3. $480 \times 1.1 = 528$ **Ans.** 528 students
4. $600 \times 1.3 = 780$ **Ans.** $780
5. $130 \times 1.1 = 143$ **Ans.** $143
6. $800 \times 0.8 = 640$ **Ans.** $640
7. $480 \times 0.8 = 384$ **Ans.** 384 people
8. $28 \times 0.75 = 21$ **Ans.** $21
9. $600 \times 0.35 = 210$ **Ans.** 210 students
10. $600 \times 0.68 = 408$ **Ans.** 408 m²

30 Graphs, Percentages & Ratios
pp 60,61

1. (1) 40 %   (2) 30 %   (3) 20 %   (4) 10 %
 (5) 2 times (double)   (6) $\frac{1}{3}$ times
2. (1) 60 %   (2) 20 %   (3) 15 %   (4) 5 %   (5) 4 times
 (6) $\frac{1}{3}$ times
3. (1) 46 %   (2) 33 %   (3) 6 %   (4) 3 %
 (5) 11 times
4. (1) 42 %   (2) 26 %   (3) 13 %   (4) 19 %   (5) $\frac{1}{2}$

31 Graphs, Percentages & Ratios
pp 62,63

1. (1) 60 %   (2) 20 %   (3) 10 %   (4) 10 %   (5) 2 times
 (6) $\frac{1}{2}$ times
2. (1) 55 %   (2) 30 %   (3) 10 %   (4) 5 %   (5) 6 times
 (6) $\frac{1}{3}$ times
3. (1) 47 %   (2) 24 %   (3) 12 %   (4) 17 %   (5) 2 times
4. (1) 48 %   (2) 24 %   (3) 12 %   (4) 8 %   (5) $\frac{1}{6}$ times

32 Graphs, Percentages & Ratios
pp 64,65

1. (1) Zone A $22 \div 50 \times 100 = 44$ %
 Zone B $13 \div 50 \times 100 = 26$ %
 Zone C $8 \div 50 \times 100 = 16$ %
 Zone D $7 \div 50 \times 100 = 14$ %
 (2) 100 %
 (3) (A) 44, (B) 26, (C) 16, (D) 14, (Total) 100
2. (1) (Roller coaster) 45, (Bumper cars) 30,
 (Merry-Go-Round) 15, (Parachute) 10, (Total) 100
 (2) (A) 34, (B) 32, (C) 20, (D) 14, (Total) 100
3. (1) (Fields) $16 \div 40 \times 100 = 40$ %
 (Gym) $14 \div 40 \times 100 = 35$ %
 (Cafeteria) $6 \div 40 \times 100 = 15$ %
 (Class) $4 \div 40 \times 100 = 10$ %
 (2) 100 %
 (3) (Fields) 40, (Gym) 35, (Cafeteria) 15, (Class) 10,
 (Total) 100
4. (1) (Coffee) 38, (Tea) 26, (Juice) 21, (Cocoa) 15,
 (Total) 100
 (2) (Comic book) 40, (Novel) 30, (Non-Fiction) 18,
 (Art book) 8, (Magazine) 4, (Total) 100

Also, number of circles on the top row is the same as the card number.
So, on the 20th card, there are
$20+21+22+23=86$ circles.

33 Graphs, Percentages & Ratios pp 66, 67

1 (1)

Residential	Agricultural	Forest	Other

0 10 20 30 40 50 60 70 80 90 100%

(2)

Steel	Chemical	Plastic	Other

0 10 20 30 40 50 60 70 80 90 100%

(3)

Protein	Carbohydrate	Water	Fat	Other

0 10 20 30 40 50 60 70 80 90 100%

2 (1) (Novel) 36, (Science) 32, (Biography) 21,
(Other) 11, (Total) 100

(2)

Novel	Science	Biography	Other

0 10 20 30 40 50 60 70 80 90 100%

3 (1)

Other, Taxi, Car, Truck

(2) Other, Books, Stationery, Snacks

(3)

Other, Precision instruments, Steel, Vehicles, Pharmaceuticals

4 (1) (Clothing) 32,
(Grocery) 26,
(Hardware) 18,
(Funiture) 3,
(Other) 21,
(Total) 100,

(2)

Other, Funiture, Hardware, Clothing, Grocery

34 Mixed Problems pp 68, 69

1 (1) 16 Coins
(2) $3 \times \boxed{4} - \boxed{4}$, $4 \times \boxed{4} - \boxed{4}$, $5 \times \boxed{4} - \boxed{4}$
(3) $20 \times 4 - 4 = 76$ **Ans.** 76 coins
2 $20 \times 3 - 3 = 57$ **Ans.** 57 coins
3 $10 + 4 \times 19 = 86$ **Ans.** 86 circles

Advice
Card #1 has 10 circles.
On card #2, there are 4 more circles.
So, on the 20th card, there are
$10 + 4 \times 19 = 86$ circles.

4 (1) $(3 + \boxed{1}) \times \boxed{3} \div \boxed{2}$, $(4 + \boxed{1}) \times \boxed{4} \div \boxed{2}$
(2) $(20 + 1) \times 20 \div 2 = 210$ **Ans.** 210 coins
5 $(20 + 1) \times 20 \div 2 = 210$ **Ans.** 210 square boards

35 Review pp 70, 71

1 $180 \times 0.8 = 144$ **Ans.** 144¢
2 $32.4 \div 1.8 = 18$ **Ans.** 18 bottles
3 $\frac{5}{7} + \frac{2}{7} = 1$ **Ans.** 1 L
4 $9 \div 8 = 1.125$ **Ans.** 1.125 times
5 $500 \times 0.2 = 100$ **Ans.** 100 notebooks
6 $3.2 \times 1.25 = 4$ **Ans.** $4
7 $\frac{5}{7} - \frac{3}{7} = \frac{2}{7}$ **Ans.** $\frac{2}{7}$ L
8 $2.5 \times 30 = 75$ **Ans.** 75 km
9 $5.6 \div 3 = 1.8\overset{9}{\cancel{6}}$ **Ans.** 1.9 m
10 $2.3 \times 4.5 = 10.35$ **Ans.** 10.35 oz.

36 Review pp 72, 73

1 $1.5 \div 1.2 = 1.25$ **Ans.** 1.25 kg
2 $2.8 \div 3.5 = 0.8$ **Ans.** 0.8 km
3 $6 \div 20 = 0.3$ **Ans.** 0.3
4 $9 \div 0.6 = 15$ **Ans.** 15 km
5 **Ans.** Black
6 $\frac{7}{8} - \frac{4}{8} = \frac{3}{8}$

Ans. The train station is $\frac{3}{8}$ km further away.

7 $50 \div 1.6 = 31$ R 0.4 **Ans.** 31 sections, 0.4 m remains
8 $29,000 - 26,000 = 3,000$ **Ans.** 3,000 people
9 $28 \times 0.75 = 21$ **Ans.** $21
10 $36 \div (3 + 1) = 9$, $36 - 9 = 27$

Ans. Red: 9 sheets, Blue: 27 sheets

[Also, $9 \times 3 = 27$]

© Kumon Publishing Co., Ltd.